FLUID MECHANICS

AUTHOR(s)

Dr. Shreenivas N. Londhe
Professor
Department of Civil Engineering
Vishwakarma Institute of Information Technology,
Pune, Maharashtra (India)

Dr. Pradnya R. Dixit
Associate Professor
Department of Civil Engineering
Vishwakarma Institute of Information Technology,
Pune, Maharashtra (India)

REVIEWER

Prof. Tanmay Basak
Professor, Department of Chemical Engineering
Indian Institute of Technology, Madras
Chennai (India)

KHANNA BOOK PUBLISHING CO. (P) Ltd.
4C/4344, Ansari Road, Darya Ganj,
New Delhi, India

All India Council for Technical Education
Nelson Mandela Marg, Vasant Kunj,
New Delhi, India

Dear Readers,

To prevent the piracy, this book is secured with HIGH SECURITY HOLOGRAM on the front title cover. In case you don't find the hologram on the front cover title, please write us to at contact@khannabooks.com or whatsapp us at +91-99109 09320 and avail special gift voucher for yourself.

Specimen of Hologram on front Cover title:

Moreover, there is a SPECIAL DISCOUNT COUPON for you with EVERY HOLOGRAM.

How to avail this SPECIAL DISCOUNT:
Step 1: Scratch the hologram
Step 2: Under the scratch area, your "coupon code" is available
Step 3: Logon to www.khannabooks.com
Step 4: Use your "coupon code" in the shopping cart and get your copy at a special discount
Step 5: Enjoy your reading!

ISBN: 978-93-5538-811-7
Book Code: UG426EN

Fluid Mechanics
by Shreenivas N. Londhe, Pradnya R. Dixit

First Edition: 2026

Originally Published by:
All India Council for Technical Education, Ministry of Education, GOI

Exclusively Published & Distributed by:
Khanna Book Publishing Co. (P) Ltd.
4C/4344, Ansari Road, Daryaganj,
New Delhi (INDIA)

Visit us at: **www.khannabooks.com**
Write us at: **contact@khannabooks.com**

To view complete list of books, Please scan the QR Code:	KPH

Printed in Bharat.

प्रो. टी. जी. सीताराम
अध्यक्ष
Prof. T. G. Sitharam
Chairman

सत्यमेव जयते

आज़ादी का
अमृत महोत्सव

अखिल भारतीय तकनीकी शिक्षा परिषद्
(भारत सरकार का एक सांविधिक निकाय)
(शिक्षा मंत्रालय, भारत सरकार)
नेल्सन मंडेला मार्ग, वसंत कुंज, नई दिल्ली–110070
दूरभाष : 011–26131498
ई–मेल : chairman@aicte-india.org

ALL INDIA COUNCIL FOR TECHNICAL EDUCATION
(A STATUTORY BODY OF THE GOVT. OF INDIA)
(Ministry of Education, Govt. of India)
Nelson Mandela Marg, Vasant Kunj, New Delhi-110070
Phone : 011-26131498
E-mail : chairman@aicte-india.org

FOREWORD

Engineers are the backbone of any modern society. They are the ones responsible for the marvels as well as the improved quality of life across the world. Engineers have driven humanity towards greater heights in a more evolved and unprecedented manner.

The All India Council for Technical Education (AICTE), have spared no efforts towards the strengthening of the technical education in the country. AICTE is always committed towards promoting quality Technical Education to make India a modern developed nation emphasizing on the overall welfare of mankind.

An array of initiatives has been taken by AICTE in last decade which have been accelerated now by the National Education Policy (NEP) 2020. The implementation of NEP under the visionary leadership of Hon'ble Prime Minister of India envisages the provision for education in regional languages to all, thereby ensuring that every graduate becomes competent enough and is in a position to contribute towards the national growth and development through innovation & entrepreneurship.

One of the spheres where AICTE had been relentlessly working since past couple of years is providing high quality original technical contents at Under Graduate & Diploma level prepared and translated by eminent educators in various Indian languages to its aspirants. For students pursuing 2nd year of their Engineering education, AICTE has identified 87 books, which shall be translated into 12 Indian languages - Hindi, Tamil, Gujarati, Odia, Bengali, Kannada, Urdu, Punjabi, Telugu, Marathi, Assamese & Malayalam. In addition to the English medium, books in different Indian Languages are going to support the students to understand the concepts in their respective mother tongue.

On behalf of AICTE, I express sincere gratitude to all distinguished authors, reviewers and translators from the renowned institutions of high repute for their admirable contribution in a record span of time.

AICTE is confident that these outcomes based original contents shall help aspirants to master the subject with comprehension and greater ease.

T. G. Sitharam
(Prof. T. G. Sitharam)

Acknowledgement

The authors are grateful to the authorities of AICTE, particularly Prof. T.G. Sitharam, Chairman; Dr. Abhay Jere, Vice-Chairman; Prof. Rajiv Kumar, Member-Secretary, Dr. Ramesh Unnikrishnan, Advisor-II and Dr. Sunil Luthra, Director, Training and Learning Bureau for their planning to publish various books for the second-year civil engineering degree level students according to new education policy (NEP 2020) and outcome-based education system. We sincerely acknowledge the valuable guidance and important suggestions received from the reviewer of this book; Dr. Tanmay Basak, Professor, IIT Madras, during the preparation of the manuscript of this book.

Authors also wish to express sincere gratitude to their teacher Late Prof. R.J. Garde because of whom they could pursue their studies further in this very interesting field of Hydraulics and Water Resources Engineering. Authors are very much thankful to Prof. Kanchan Khare, Dept. of Civil Engg., Symbiosis Institute of Technology, Pune for her encouragement, suggestions, and critical comments.

Authors are grateful to Shri Bharat Agarwal; Managing Trustee of BRACT'S, Vishwakarma Institute of Information Technology (VIIT), Pune, Maharashtra, India and Prof. Vivek Deshpande, Director, VIIT, Pune; for their support.

Authors extend sincere regards to their colleague Dr. Preeti Kulkarni for her continuous support during the manuscript preparation of this book as well as for her suggestions and critics.

Authors are also thankful to their family members for their timely help and their patience when the work of this book was in progress.

Acknowledgements are due to the contributors and different workers in this field whose published books, review articles, papers, photographs, footnotes, references, and other valuable information enriched us at the time of writing the book.

Finally, authors are gratified to their each and every student because, while teaching them, they received "experience of teaching" and have enriched their knowledge of fluid mechanics time to time.

Prof. Shreenivas Londhe

Dr. Pradnya Dixit

Preface

It gives us immense pleasure to present our book on Fluid Mechanics as per AICTE curriculum of second year civil engineering degree course followed by concept of outcome based education as per National Education Policy (NEP)2020.

This book is a specially prepared student centric book in which different domains of fluid Mechanics are introduced in simple manner. This book includes fundamnetals of fluid mechanincs in total of six units as properties of fluids, fluid statics, buoyancy and flotation, fluid kinematics, fluid dynamics and dimensional analysis. Efforts have been made to explain the fundamental concepts of the subject in the simplest possible way. Main purpose of this book is to help under graduate civil engineering students to understand and apply the basics of fluid mechanics to applications in engineering problems. The content of this book is aligned with the model curriculum of AICTE by mapping of Course Outcome, Programs Outcomes and Unit Outcomes. At the start of each unit, rationale, pre-requisite of that unit are mentioned along with unit (learning) outcomes to make the students aware about the expected outcome from the same unit . In addition to the essential information, experiments related to the units are provided along with the objecties , procedres and necessary calcualtion basics. Every unit is well supported by a set of objective questions, theory questions and numerical problems in addition to the solved numerical examples. The subject matters are presented in a constructive manner so that an engineering degree prepares students to work in different sectors or in national laboratories at the very forefront of technology. We sincerely hope that the book will inspire the students to learn and discuss the ideas behind basic principles of Fluid Mechanics and will surely contribute to the development of a solid foundation of the subject. We would be thankful to all beneficial comments and suggestions which will contribute to the improvement of the future editions of the book. It gives us utmost satisfaction to place this book in the hands of the teachers and students. The authors extend best wishes to students for the preparation of the fluid Mechanics course.

Prof. Shreenivas Londhe

Dr. Pradnya Dixit

Outcome Based Education

For the implementation of an outcome-based education the first requirement is to develop an outcome-based curriculum and incorporate an outcome-based assessment in the education system. By going through outcome-based assessments evaluators will be able to evaluate whether the students have achieved the outlined standard, specific and measurable outcomes. With the proper incorporation of outcome-based education there will be a definite commitment to achieve a minimum standard for all learners without giving up at any level.
At the end of the programme running with the aid of outcome-based education, a student will be able to arrive at the following outcomes:

PO1. Engineering knowledge: Apply the knowledge of mathematics, science, engineering fundamentals, and an engineering specialization to the solution of complex engineering problems.

PO2. Problem analysis: Identify, formulate, review research literature, and analyze complex engineering problems reaching substantiated conclusions using first principles of mathematics, natural sciences, and engineering sciences.

PO3. Design / development of solutions: Design solutions for complex engineering problems and design system components or processes that meet the specified needs with appropriate consideration for the public health and safety, and the cultural, societal, and environmental considerations.

PO4. Conduct investigations of complex problems: Use research-based knowledge and research methods including design of experiments, analysis and interpretation of data, and synthesis of the information to provide valid conclusions.

PO5. Modern tool usage: Create, select, and apply appropriate techniques, resources, and modern engineering and IT tools including prediction and modeling to complex engineering activities with an understanding of the limitations.

PO6. The engineer and society: Apply reasoning informed by the contextual knowledge to assess societal, health, safety, legal and cultural issues and the consequent responsibilities relevant to the professional engineering practice.

PO7. Environment and sustainability: Understand the impact of the professional engineering solutions in societal and environmental contexts, and demonstrate the knowledge of, and need for sustainable development.

PO8. Ethics: Apply ethical principles and commit to professional ethics and responsibilities and norms of the engineering practice.

PO9. Individual and team work: Function effectively as an individual, and as a member or leader in diverse teams, and in multidisciplinary settings.

PO10. Communication: Communicate effectively on complex engineering activities with the engineering community and with society at large, such as, being able to comprehend and write effective reports and design documentation, make effective presentations, and give and receive clear instructions.

PO11. Project management and finance: Demonstrate knowledge and understanding of the engineering and management principles and apply these to one's own work, as a member and leader in a team, to manage projects and in multidisciplinary environments.

PO12. Life-long learning: Recognize the need for, and have the preparation and ability to engage in independent and life-long learning in the broadest context of technological change.

Course Outcomes

At the end of the course, the student will be able to:

CO-1: Understand the broad principles of fluid statics, kinematics and dynamics

CO-2: Understand definitions of the basic terms used in fluid mechanics

CO-3: Understand classifications of fluid flow

CO-4: Apply the continuity, momentum, and energy principles

CO-5: Apply dimensional analysis

Course Outcomes	Expected Mapping with programme outcomes *(1: Weak Correlation; 2: Medium Correlation; 3: Strong Correlation)*												
	PO-1	PO-2	PO-3	PO-4	PO-5	PO-6	PO-6	PO-7	PO-8	PO-9	PO-10	PO-11	PO-12
CO-1	3	1	1	1	1	-	-	-	-	-	-	-	-
CO-2	3	-	1	-	1	-	-	-	-	-	-	-	-
CO-3	3	1	1	1	1	-	-	-	-	-	-	-	-
CO-4	3	1	1	1	1	-	-	-	-	-	-	-	-
CO-5	3	2	1	1	1	-	-	-	-	-	-	-	-

Guidelines for Teachers

To implement Outcome Based Education (OBE) knowledge level and skill set of the students should be enhanced. Teachers should take a major responsibility for the proper implementation of OBE. Some of the responsibilities (not limited to) for the teachers in OBE system may be as follows:

- Within reasonable constraint, they should manoeuvre time to the best advantage of all students.
- They should assess the students only upon certain defined criterion without considering any other potential ineligibility to discriminate them.
- They should try to grow the learning abilities of the students to a certain level before they leave the institute.
- They should try to ensure that all the students are equipped with the quality knowledge as well as competence after they finish their education.
- They should always encourage the students to develop their ultimate performance capabilities.
- They should facilitate and encourage group work and team work to consolidate newer approach.
- They should follow Blooms taxonomy in every part of the assessment.

Bloom's Taxonomy

Level	Teacher should Check	Student should be able to	Possible Mode of Assessment
Create	Students' ability to create	Design or Create	Mini project
Evaluate	Students' ability to justify	Argue or Defend	Assignment
Analyze	Students' ability to distinguish	Differentiate or Distinguish	Project/Lab Methodology
Apply	Students' ability to use information	Operate or Demonstrate	Technical Presentation/ Demonstration
Understand	Students' ability to explain the ideas	Explain or Classify	Presentation / Seminar
Remember	Students' ability to recall (or remember)	Define or Recall	Quiz

Guidelines for Students

Students should take equal responsibility for implementing the OBE. Some of the responsibilities (not limited to) for the students in OBE system are as follows:

- Students should be well aware of each Unit Outcome (UO) before the start of a unit in each and every course.
- Students should be well aware of each Course Outcome (CO) before the start of the course.
- Students should be well aware of each Programme Outcome (PO) before the start of the programme.
- Students should think critically and reasonably with proper reflection and action.
- Learning of the students should be connected and integrated with practical and real-life consequences.
- Students should be well aware of their competency at every level of OBE.

Abbreviations and Symbols

List of Abbreviations

General Terms

Abbreviations	Full form
SI	System International
MKS	Meter kilogram second
CGS	Centimeter gram second

List of Symbols

Symbol	Description
A, a	Area
a	acceleration
b	width
B	centre of buoyancy
c	Celerity of pressure wave (acoustic velocity)
C.P	centre of pressure
C_c	coefficient of contraction
C_D	coefficient of drag
CG, G	centre of gravity
C_L	coefficient of lift
C_V	coefficient of velocity
D (d)	Diameter
dp	Change in pressure
dv	Change in volume
E	specific energy
Eu	Euler number
f	friction factor (Darcy) for pipe flow
F	force, thrust

Fe	elastic force
Fg	buoyant force
Fi	Inertia force
Fp	Pressure force
Fr	Froude number
Fs	Viscus force
Fs	surface tension force
g	gravitational acceleration
g	Acceleration due to gravity
h	head, height or depth, pressure head
H	total head (energy)
hf	frictional head losses
hp	horsepower = 0.746 kW
I	moment of inertia
I_{XY}	product of inertia
J	joule
J	joule
K	bulk modulus of elasticity,
K.E.	kinetic energy
l	mixing length
L	length
Le	equivalent length
M	Mass, metacentre
MG	metacentric height
N	Speed, Newton
Ny	Mach number
P	pressure, Power
Pa	pascal
PE	potential energy
PE	pressure energy
q	unit flow, unit discharge
Q	Discharge, rate of flow

Re	Reynold's number
rpm	rotational speed
S	Specific gravity, relative density
T	Time, torque
V	velocity, volume
Vs	Specific volume
W	weight , watt
We	Weber number
α (alpha)	angle, kinetic energy correction factor
β (beta)	angle, momentum correction factor
γ (gamma)	specific (or unit) weight
τ	shear stress
δ (delta)	boundary layer thickness
ε (epsilon)	surface roughness
η (eta)	eddy viscosity
θ	Momentum thickness
θ (theta)	any angle
μ (mue)	absolute viscosity
ν (nu)	kinematic viscosity
π (pi)	dimensionless parameter
ρ (rho)	mass density
σ (sigma)	surface tension, intensity of tensile stress
τ (tau)	shear stress
φ (phi)	speed factor, velocity potential, ratio
ψ (psi)	stream function
ω (omega)	angular velocity
$\frac{du}{dy}$	angular deformation
δ^*	Displacement thickness

List of Figures

UNIT 3

UNIT 4

List of Tables

Table of Contents

1 PROPERTIES OF FLUIDS

UNIT SPECIFICS

Through this unit we have discussed the following aspects:

- *Introduction of fluid mechanics, Liquids and Gases, Units of measurement*
- *Properties of fluids, Mass, Mass Density, Specific Weight, Specific Volume, Specific Gravity*
- *Viscosity, Kinematic and dynamic viscosity, Classification of fluids*
- *Compressibility & Bulk modulus of elasticity*
- *Cohesion and Adhesion, Surface Tension, Capillarity*
- *Vapor pressure (Boiling point, Cavitation)*

This is followed by large number of solved examples. The students are encouraged to solve the objective questions, long answer questions and numerical problems to judge ones understanding. The practical on measurement of viscosity is included followed by a list of references for additional reading.

RATIONALE

This unit introduces basic properties of fluids which are necessarily to be understood before learning the mechanics of fluids. To understand what happens to the fluid under the action of forces either at rest or in motion one must first understand the internal structure of the fluid, its classification, response to forces and behaviour with changes in physical parameters like temperature and pressure.

PRE-REQUISITES

Mathematics: Derivatives (Class XII)

Physics: Mechanics (Class XII)

UNIT OUTCOMES

List of outcomes of this unit is as follows:

(At the end of this unit, students will understand.)

U1-O1: Difference between liquids and gases as fluids

U1-O2: Units of measurement

U1-O3: Properties of fluids

U1-O4: Classification of fluids

Unit-1 Outcomes	EXPECTED MAPPING WITH COURSE OUTCOMES (1-WeakCorrelation; 2-Mediumcorrelation;3-StrongCorrelation)					
	CO-1	CO-2	CO-3	CO-4	CO-5	CO-6
U1-O1	-	3	-	-	-	-
U1-O2	-	3	-	-	-	-
U1-O3	-	3	-	-	-	-
U1-O4	-	3	-	-	-	-

1.1 Introduction:

Matter in the universe exists in three states, the solids, liquids and gases. This classification is based on spacing of different molecules of the matter and the response of the matter to the stress. The solids exhibit rigidity of form owing to the closeness of molecules in them. Their form/shape remains same in normal temperature and pressure situations. On the other hand, the liquids and gases show a very less rigidity of form i.e., they take the shape of the container. This is due to the larger distance between their molecules. The gases have more freedom of movement due to greater distance between their molecules than the distance between molecules of liquids. When subjected to a shearing force solid tend to deform which is resisted by the internal resistance (shear stress). If the force is removed solids regain their original shape and size owing to their resilience. If the force exceeds certain limit the solids deform permanently. On the other hand, the liquids and gases deform continuously under the action of shear, however small it may be and start flowing. Owing to this behaviour of 'flowing', the liquids and gases are termed as fluids. Thus, fluid can be defined as a substance which deforms continuously when subjected to a tangential or shear stress (force), however small the stress (force) may be.

Mechanics is a branch of science which deals with study of action of forces on bodies in a state of rest or in motion. When rules of mechanics are applied to fluids, it is termed as ***'Fluid Mechanics'.*** *When fluid is at rest it is called* ***'fluid statics'*** *while when the fluid is in motion it is termed as* ***'fluid dynamics****. The fluid dynamics can be studied either by applying the forces responsible for the motion (fluid dynamics) or without applying these forces (fluid kinematics) on the similar lines of solid mechanics. To understand fluid mechanics, it is first necessary to study the properties of fluids and their interrelationships which are discussed in the present unit.*

Fluid Mechanics has a wide range of applications in almost all hardcore engineering branches like Civil, Mechanical, Chemical and Electrical Engineering. The applications include design of water supply schemes, design of hydraulic structures, design of hydraulic machines, design of chemical industries, power generation, design of lubricating systems, design of aeroplanes, submarines, design of ships, the list is unending.

1.2 Liquids and Gases:

As discussed earlier the fluids are of two types-liquids and gases. These two phases have many similarities and differences. The comparison between the liquids and gases in respect of their properties is given in the following table 1.2.1

Table 1.2.1: Difference between liquids and gases

Sr. No.	Liquids	Gases
1	Do not offer resistance to shear	Do not offer resistance to shear
2	Occupy the definite volume in the container	Occupy the entire volume of the container
3	Incompressible compared to gases	Highly compressible
4	Exhibit free surface as molecules are kept together due to relatively	Do not have a free surface as molecules travel away from each
5	Liquids can easily change their phase from liquid to gas state	Gases do not convert from gas to liquid state easily (except water

1.3 Units of Measurement:

The standards used for measurement of physical quantities like length, mass, volume, acceleration etc. are called as **units.** In fluid mechanics units of measurement are based on Newton's second law viz.

$$F = ma$$

where F is force, m is mass and a is the acceleration in the direction of motion.

In absolute system the primary unit is mass where as in gravitational system force is the primary unit.

Since, 1960, the 'System International d units' is followed (SI system) internationally to measure various physical quantities. The primary or basic units are given in Table 1.3.1. Frequently used derived units are listed in Table 1.3.2.

Table1.3.1: Basic or fundamental units

Quantity	Units	Symbol
Length	Meter	M
Mass	Kilogram	Kg
Time	Second	S
Electric current	Amphere	A
Luminance	Candela	Ed
Temperature	Celsius	C
Thermodynamic Temperature	Kelvin	K

Table 1.3.2: Derived units

Quantity	Unit	Symbol
Force	Newton	N
Pressure	Pascal	Pa (N/ m^2)
Stress	Pascal	Pa (N/ m^2)
Work/energy	Joule	J (N-m)
Power	Watt	W (N m/s)

1.4 Properties of Fluid:

To study behaviour of fluids under the action of forces following properties are significant.

1.4.1 Mass:

The quantity of matter present in the system is expressed by mass. Mass is an indication of the amount of effort required to start or stop the motion, as effort required is directly proportional to mass.

Unit of measurement: kg

1.4.2 Mass Density:

It is a measure of compactness of the matter present in a fluid. The mass density is expressed as mass per unit volume of the fluid denoted by symbol 'ρ' (Rho). Thus, more mass density means more matter present per unit volume.

$$\rho = \frac{\text{Mass}}{\text{Volume}}$$

Unit of measurement: kg/m^3

The mass density depends upon the temperature and pressure as volume is a function of Equation should be centrally placed, it is right justified at present temperature and function. The standard value of mass density of water at 20^0 C and 1 atmospheric pressure is 998 kg/m^3 and that of air is 1.205 kg/m^3.

1.4.3 Specific Weight:

It is weight of the fluid per unit volume. It is denoted by symbol gamma (γ).

$$\gamma = \frac{\text{Weight}}{\text{Volume}} = \frac{\text{mass x gravitational acceleration}}{\text{Volume}}$$

i.e.

$$\gamma = \frac{W}{V} = \frac{m \text{ x } g}{V} = \rho \, . \, g$$

$$\frac{m \text{ x } g}{V} = \frac{kg \, . \, m/s^2}{m^3} = \frac{N}{m^3}$$

Unit of measurement: N /m^3

Specific weight of water at 20^0 C is 9810 N/m^3 and that of air is 11.81 N/m^3. The specific weight is the product of mass density and gravitational acceleration ($\gamma = \rho.g$)

1.4.4 Specific Volume:

It is inverse of specific weight i.e., it is the volume of fluid per unit weight. It is denoted by symbol 'V_s.'

$$V_s = \frac{1}{\gamma} = \frac{\text{Volume}}{\text{Weight}}$$

Specific volume is used in problems involving gas flows.
Unit of measurement: m^3/N

1.4.5 Specific Gravity:

It is the ratio of specific weight of fluid to the specific weight of standard fluid at standard conditions. For liquids, water at 4^0C temperature and 101.325 kN/m^2 pressure is used as a standard fluid. For gases carbon dioxide or pure hydrogen at standard temperature or pressure is used as a standard fluid. It is denoted by symbol 'S'.

$$\text{Specific gravity } 'S' = \frac{\text{Specific weight of fluid}}{\text{Specific weight of standard fluid}}$$

Thus, specific gravity of water will be l. Specific gravity of Mercury is 13.6. If instead of specific weight, mass density is used to find ratio, then the resulting term is known as relative density (R.D.)

$$\therefore \; R.D. \; = relative\ density \; = \frac{mass\ density\ of\ fluid}{mass\ density\ of\ standard\ fluid}$$

Both specific gravity and relative density are dimensionless and possess no 'unit'.

1.4.6 Viscosity:

Consider a case, when water is dropped on floor and oil is dropped on floor. It can be observed that water flows easily and quickly than oil. This is due to the influence of a property called viscosity. It can be easily understood from the above example that the resistance to flow in oil is greater than that of water. Viscosity is the measure of this resistance to flow.

Viscosity is the property of fluid by virtue of which it offers resistance to movement of one layer of the fluid over the other.

Viscosity of a fluid is mainly due to intermolecular cohesion and molecular momentum. While the former is dominant in liquids the latter is dominant in gases. The increase in temperature results in reduction in intermolecular cohesion due to increased spacing between liquid molecules leading to reduction in the viscosity. However, in gases the intermolecular momentum increases with temperature, thereby increasing viscosity with temperature.

Consider two parallel plates separated by a distance Y apart. The space between the plates is filled with fluid. The upper plate is moving with velocity 'U' due to application of force F. (Refer Fig. 1.l). The lower plate is stationary.

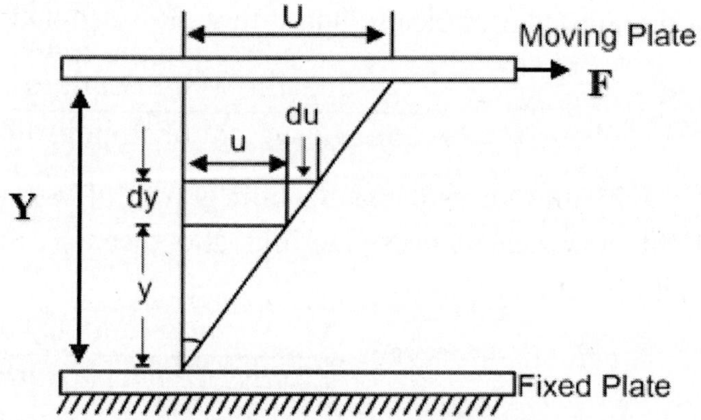

Fig. 1.1 Viscosity of fluid

Fluid particles in contact with the plate have same velocity as that of the plate which is termed as the 'No slip condition'. Thus, the velocity of fluid particles varies from zero at the stationary plate to 'U' at the moving plate.

Consider fluid layers at a distance 'y' and 'y + dy' from the stationary plate. The velocity of the flow at these levels is 'u' and 'u + du' respectively setting up a velocity gradient of $\frac{du}{dy}$. This happens due to the shear resistance offered by both the plates to the relative motion between them. According Newton's law of viscosity the shear stress between two straight parallel lines of non-turbulent flow is proportional to the rate of change of velocity with respect to y (velocity gradient).

$$\tau \propto \frac{du}{dy} \qquad (1.1)$$

$$\therefore \quad \tau = \mu \frac{du}{dy} \qquad (1.2)$$

Where τ is the shear stress, $\frac{du}{dy}$ is the velocity gradient and μ (mue), the constant of proportionality, is termed as coefficient of dynamic viscosity. The law stated by equation (1.2), is known as Newton's law of viscosity. The force acting on the plate can be expressed as,

$$F = \tau \,.\, A \qquad \text{where A is area of the plate} \qquad (1.3)$$

$$\therefore F = \mu \frac{du}{dy} A \qquad (1.4)$$

Rearranging the terms of Equation (1 .2)

$$\mu = \frac{\tau}{du/dy} \qquad (1.5)$$

where $\frac{du}{dy}$ can be also considered as angular deformation.

The coefficient of dynamic viscosity can therefore be defined as the shear stress required to produce a unit angular deformation.

Unit of viscosity μ is N-s /m^2 or Pa-s or kg/ms.

In C.G.S, system unit of viscosity is poise i.e. dyne-s / cm^2

$$1 \text{ poise} = 0.1 \text{ Ns/m}^2$$

$$1 \text{ centipoise} = \frac{1}{100} \text{ poise} = 0.01 \text{ poise}$$

The ratio of dynamic viscosity 'μ' to mass density 'p' is termed as coefficient of kinematic viscosity 'v' (nue).

$$\nu = \frac{\mu}{\rho} \tag{1.6}$$

Unit of measurement: m^2/s

In C.G.S. system stokes i.e. cm^2/s is used to measure kinematic viscosity

$$1 \text{ stokes} = 10^{-4} \text{ m}^2/\text{s}$$

$$1 \text{ centistoke} = \frac{1}{100} \text{stokes}$$

The dynamic viscosity depends on the temperature on account of the density which in turn depends upon the temperature. However, the kinematics viscosity is independent of the temperature as it does not involve the density as seen in equation 1.6. When fluids are at rest there is no relative movement between lateral layers, as a result there is no shear stress. Therefore, viscosity is not important in study of fluids at rest.

The fluids which possess viscosity are called real fluids. For example, milk, blood, oil, water etc. The fluids which do not possess viscosity are Ideal fluids. As the name indicates these fluids are Ideal and therefore such fluids do not exist in reality. Only if viscosity is very less then effect of viscosity can be neglected and the flow can be considered as inviscid flow. The property is viscosity can be further used to classify the fluids.

1.4.6.1. Classification of Fluids:

Depending upon the property of viscosity and Newton's law of viscosity the fluids can be classified as under-

Newtonian fluids:

These are the fluids which obey Newton's law of viscosity which means the constant of proportionality μ (coefficient of dynamic viscosity) remains constant. Relationship between shear stress and angular deformation is a straight line.

Example: air, water, kerosene, glycerine etc.

Non-Newtonian fluids:

These fluids do not obey Newton's law of viscosity. Non-Newtonian fluids are further classified depending upon the yield stress.

Those non-Newtonian fluid which does not possess a yield stress are governed by the following non-linear relationship between the shear stress and angular deformation.

$$\tau = \mu \left(\frac{du}{dy}\right)^n \tag{1.7}$$

If n < 1, the fluids are called **'pseudoplastics'.** Liquids such as milk, blood, paper pulp behave as pseudoplastics.

If n > 1, the fluids are called **'Dilatants'.** At low shear stress they behave as fluids but at high values of shear they behave as solids. Concentrated sugar solution, butter are examples of dilatants.

Some non-Newtonian fluids possess a definite yield stress beyond which the shear stress varies linearly with angular deformation i.e.

$$\tau = \text{constant} + \mu \left(\frac{du}{dy}\right) \tag{1.8}$$

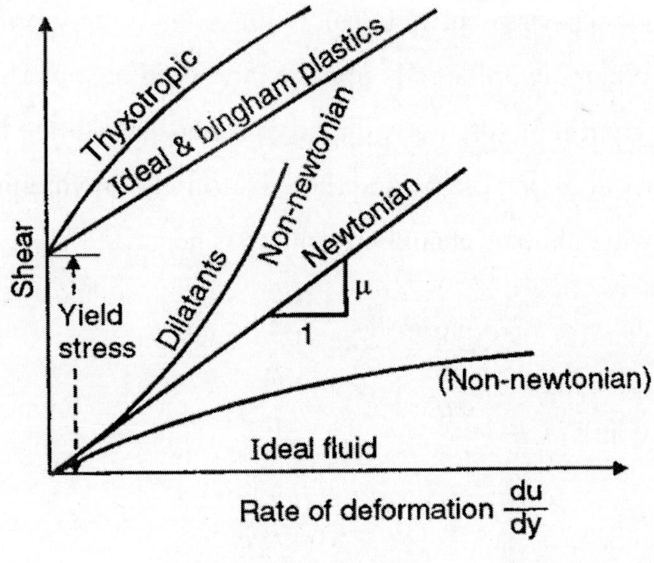

Fig. 1.2 Classification of fluids

These fluids are called as **ideal plastics or Bingham plastics.** Sewage sludge, tooth paste, oil paints, drilling mud are examples of ideal plastics.

Those fluids which possess a definite yield stress while relationship between shear stress and angular deformation is nonlinear are called as **Thixotropic fluids**.

$$\tau = \text{constant} + \mu \left(\frac{du}{dy}\right)^n \tag{1.9}$$

Printer's ink and lipsticks fall under this category.

The diagrammatic representation of classification of fluids is shown in Fig.1.2 where angular deformation (shear strain) is plotted on the X-axis and shear stress is plotted on the Y-axis. Thus, X-axis also represents the ideal fluids having zero shear stress and Y-axis also represents elastic solids having no angular deformation.

1.4.7 Compressibility:

It is the measure of elasticity in fluids. Fluids are compressed under pressure due to change in their mass density i.e. more mass can be accommodated in the unit volume. When the pressure is removed the fluid regains its original volume. The change in the pressure is directly proportional to the ratio of change in volume per original volume and there is decrease in volume with increase in pressure which is why change in volume is negative (-dV).

$$dp \ \alpha \ -\frac{dV}{V}$$

$$dp = K\,(-\frac{dV}{V})$$

$$\therefore \quad K = -\frac{dp}{(dV/V)} \tag{1.10}$$

where the constant of proportionality K is known as bulk modulus of elasticity. Thus, if the change in the volume is less, less is the compressibility (dV/V) while the bulk modulus of elasticity K is more and vice versa. The compressibility is therefore expressed as inverse of bulk modulus. Higher the bulk modulus less is the compressibility of fluids.

The liquids are considered to be incompressible owing to negligible change in the mass density at a given temperature under the action of pressure.

On the other hand, gases are treated as highly compressible.

At normal temperature and pressure

$$K_{water} = 2.07 \times 10^6 \text{ kPa}$$

$$K_{air} = 101.3 \text{ kPa}$$

$$K_{ms} = 2.07 \times 10^8 \text{kPa}$$

These values show that air is 20,000 times compressible than water and water is 100 times compressible than solid (Mild Steel)). Hence, in problems like water hammer, water is considered as compressible however for majority of the fluid mechanics problems water and other liquids are treated as incompressible

The property of compressibility plays a very important role in aerodynamics.

1.4.8 Cohesion and Adhesion:

Cohesion:

Molecular attraction between molecules of similar types (of same liquid) is termed as cohesion.

Adhesion:

Molecular attraction between dissimilar type of molecules is called as adhesion. For example: water molecules and our body.

A liquid may have both cohesion and adhesion or only cohesion without adhesion (Mercury). These properties of cohesion and adhesion lead towards two important phenomena namely surface tension and capillarity.

1.4.8.1 Surface Tension:

Consider a molecule 'x' (Refer Fig.1.3) well below the free water surface and another molecule 'y' near the free water surface.

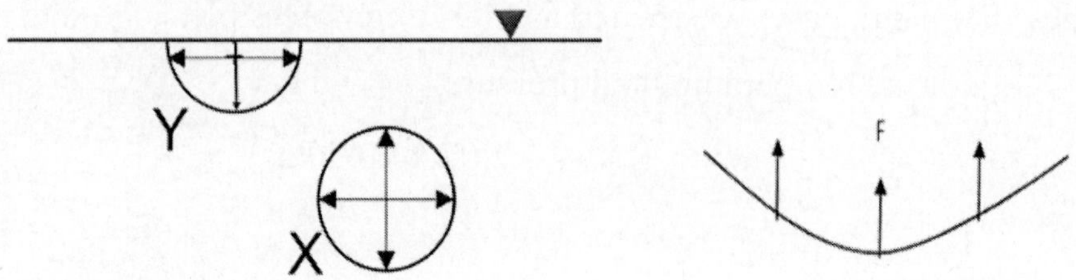

Fig. 1.3 Intermolecular forces near liquid surface

The molecule 'x' will be stable as it will be attracted equally by cohesive forces of neighbouring water molecules from all the directions. However, 'y' will always experience resultant unbalanced downward force (F) acting perpendicular to the surface as it will not experience the cohesive force of water molecules from top due to interface of water and air.

As a result, the surface will be pulled down showing a curvature. This in turn develops a tension (F) in the surface (as shown in figure. This property of liquid to exert tension is called as surface tension.

Surface tension is a result of pressure difference created at the interface of two fluids, due to cohesive forces of different magnitude acting on the molecules near the free surface. The surface tension coefficient (σ), therefore is defined as the force per unit length required to hold that surface together at that line.

$$\text{Coefficient of surface tension } \sigma = \frac{F}{L} \tag{1.11}$$

$$\therefore \text{ unit} = \text{N/m}$$

The surface tension depends on temperature, pressure and the substance it is in contact with.

$$\sigma_{water} \text{ at } 200^\circ \text{ C} = 0.0736 \text{ N/m}$$

$$\sigma_{water} \text{ at } 1000° \text{ C} = 0.0589 \text{ N/m}$$

Once the surface becomes curved, the existence of normal force requires an equal and opposite force to maintain the static equilibrium. The raindrop, soap bubble, jet of liquid thus develops an excess pressure inside, thereby creating a pressure difference to balance surface tension

(i) Pressure intensity inside a droplet:

Consider a spherical droplet of diameter 'd'. Let the excess pressure developed inside droplet is P. Figure.1.4 shows the free body diagram of one half of droplet. The forces acting are (i) Surface tension acting on the circumference. (σ), (ii) Excess pressure 'P' inside the bubble acting on area.

Surface tension force = σ x π x d.

Force due to excess pressure $= p \times \frac{\pi}{4} \times d^2$

For equilibrium, $\sigma \times \pi \times d = p \times \frac{\pi}{4} \times d^2$

$$\therefore\ p = \frac{4\sigma}{d} \qquad (1.12)$$

Thus, pressure intensity inside a droplet varies inversely with the diameter.

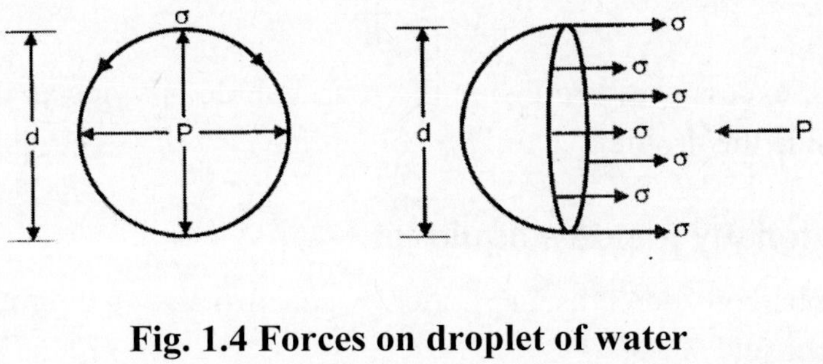

Fig. 1.4 Forces on droplet of water

(ii) Pressure intensity inside a soap bubble:

Unlike droplet, soap bubble has two surfaces in contact with air, one inside and one outside because of the small thickness. Fig.1.5 shows the free body diagram of half of soap bubble. Thickness of soap bubble is δd.

Force due to excess pressure = $p\,.\frac{\pi}{4}(d - \delta.d)^2$

$= p\,.\frac{\pi}{4}d^2$ (neglecting δd term as it is very small)

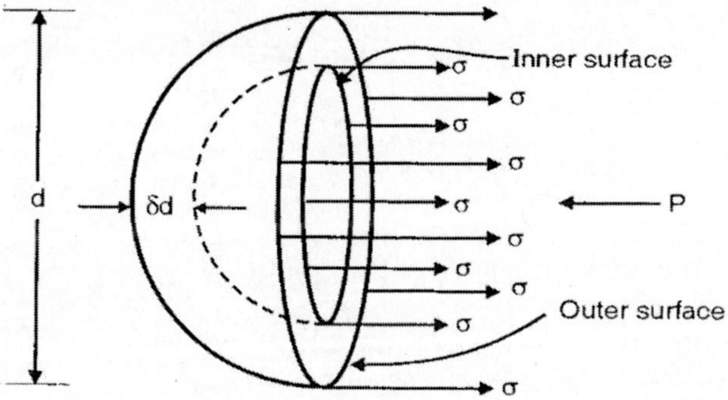

Fig. 1.5 Forces on bubble

Forces due to surface tension = $\sigma\,.\pi\,.d + \sigma\,.\pi\,.\,(d - \delta d) = 2\,\sigma\pi d$ (neglecting δd)

For equilibrium, $p\,.\frac{\pi}{4}d^2 = 2\,\sigma\pi d$

$$\therefore\ \text{p} = \frac{8\sigma}{\text{d}} \qquad (1.13)$$

Thus, excess pressure inside the soap bubble is twice as that of the pressure intensity inside the droplet.

(iii) Pressure intensity inside a liquid jet:

The liquid jet can be considered as a cylinder. Figure 1.6 shows the free body diagram of one half of the jet.

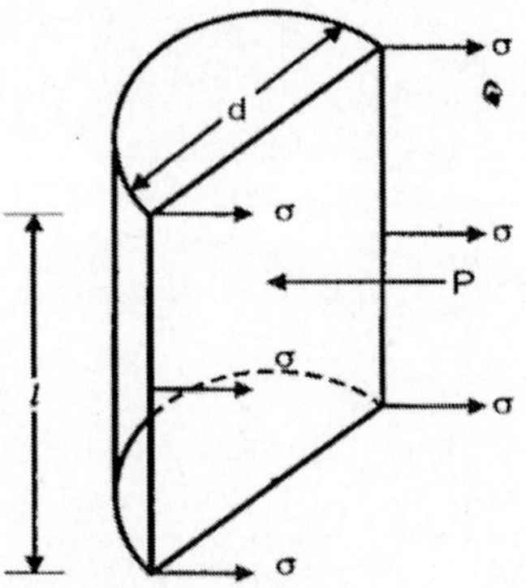

Fig. 1.6 Forces on jet liquid

Force due to excess pressure = p x d x l

Force due to surface tension = σ x 2 x l

For equilibrium, p x d x l = σ x 2 x l

$$\therefore\ p = \frac{2\sigma}{d} \qquad (1.14)$$

1.4.8.2 Capillarity:

One of the important applications of cohesion and adhesion is capillary rise or capillarity depression in small diameter tubes and interstices of porous material. This can be seen by a simple experiment. Immerse a small diameter tube (< 6 mm) vertically inside a pool of a liquid. The liquid level inside the tube will either rise or fall compared to general liquid level in the pool. This rise or fall of liquid level inside the tube is known as capillary rise or fall. If adhesive forces predominate, as in case of water, the liquid will wet the glass surface and the liquid level will rise making the liquid surface in the tube (meniscus) concave upwards (Figure:1.7 (a)).

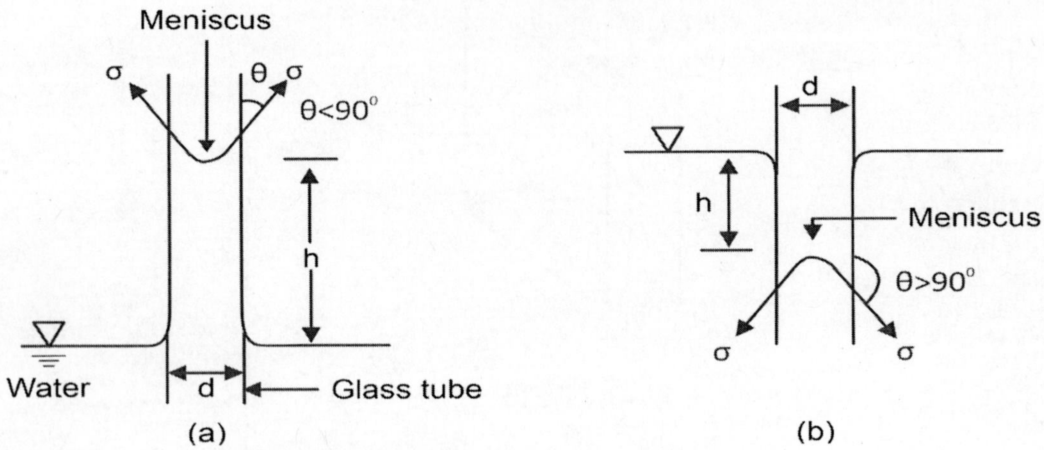

Fig. 1.7 Capillarity rise and depression

In such a situation the angle made by surface tension with the vertical, θ (theta) is less than 90^0.If cohesive forces predominate (mercury) the liquid level inside the tube (meniscus)will become convex upwards. In this situation, the angle 'θ' is greater than 90^0. (Figure:1.7 (b)).

Let 'h' be the capillary rise or fall in a tube of diameter 'd'. Under equilibrium the weight of liquid column will be balanced by vertical component of surface tension force 'σ cos θ'.

weight of liquid column = Volume x specific weight $=\left[\frac{\pi}{4} d^2 \ h\right]\gamma$

where 'γ' is specific weight of liquid.

Surface tension force = (π .d).σ cos θ

$\left[\frac{\pi}{4} d^2 \ h\right]\gamma = (\pi .d)\ \sigma \cos\theta$

$$\therefore h = \frac{4\,\sigma\cos\theta}{\gamma\, d} \tag{1.15}$$

For pure water in contact with clean glass and air, θ = 0

$$\therefore h = \frac{4\,\sigma}{\gamma_w\ d} = \frac{\sigma}{\gamma_w\ r}$$

For Mercury and glass, θ = 140^0

In the above derivation it is assumed that meniscus or curved liquid surface is section of a sphere with radius less than 2.5 mm with liquid and tube surfaces extremely clean. Therefore actual 'h' is generally less than calculated value.

For glass tubes having diameter more than 6 mm, effect of capillarity is negligible due to gravity forces become more appreciable and meniscus becoming less spherical in shape. Therefore, for pressure measurement the glass tubes are of diameter more than 6 mm.

1.4.9 Vapour Pressure:

All liquids tend to evaporate when exposed to atmosphere.

Consider a closed container partly filled with a liquid and maintained at a constant temperature. The molecules at the free surface escape into the space above the free surface.

Some of the molecules come back and some do not. After a while a stage is reached when the number of molecules on liquid surface remains constant due to the equal rate of molecules entering and leaving the liquid. The partial pressure exerted by these vapour molecules on water surface is termed as vapour pressure or saturated vapour pressure. The pressure exerted by other gases on the liquid surface along with the vapour pressure is the total pressure acting on the liquid. Vapour pressure depends upon temperature i.e. vapour pressure increase with temperature and vice-versa. When the pressure above the liquid is equal to or less than the vapour pressure liquid starts boiling. The atmospheric pressure on water is 10^5Pa and vapour pressure on free water surface is approximately 10^5Pa at 100^0C, as a result water starts boiling at 100^0 C.

Thus, boiling point of water can be decreased by increasing the vapour pressure. This underlines the working principle of a pressure cooker.

It is also established that higher the vapour pressure more volatile is the liquid. For example, petrol has a vapour pressure of 30400 Pa at 20^0C whereas Mercury has 0.16 Pa. Therefore, Mercury is used as a manometric liquid and not petrol. Due to vaporization of liquids pockets of dissolved gases and vapours are formed which are carried away by the flowing liquid in the high-pressure region where they collapse giving rise to high impact pressure.

This phenomenon is known as **cavitation** which is disastrous due to its destructive nature as material from adjoining structures is eroded. Therefore, every attempt is made in design of hydraulic structures to maintain total pressure over the liquids well above the vapour pressure.

QR CODES FOR SUPPORTING VIDEO LINKS

(1) (2) (3) (4)

1.5 Solved Problems:

Ex.1.1 Eight litres of liquid of specific gravity 0.8 is mixed with ten litre liquid of specific gravity 1.3. If the bulk of the combined liquid shrinks by one percent on mixing, determine the specific gravity, the volume and the weight of the mixture.

Solution:

Specific weight of 1stliquid = 0.8 x 9810 = 7848 N/m^3

Volume of 1stliquid = 8 lit = 0.008 m^3

$\therefore$ Weight of 1st liquid = 7848 x 0.008 = 62.784 N

Specific weight of 2nd liquid = 1.3 x 9810 = 12753 N /m^3.

Volume of 2ndliquid = 10 lit = 0.01 m^3

$\therefore$Weight of 2ndliquid = 12753 x 0.01 = 127.53 N

Reduction in the volume of mixture = $\frac{1}{100}(0.008 + 0.01) = 1.8 \text{ x}10^{-4}\text{m}^3$

	$\therefore$Volume of mixture = $(0.008 + 0.01) - 1.8\,x10^{-4} = 0.01782\ m^3$ Weight of mixture = 62.784 + 127.53 = 190.314 N Specific weight of mixture = $\frac{190.314}{0.01782} = 10679.8\ N / m^3$ Specific gravity of mixture = $\frac{10679.8}{9810} = 1.088$

Ex.1.2	Shear stress at a point is 0.6 Pa where velocity gradient is 1.5/s. If kinematic viscosity of the flowing liquid is 4.65 stokes determine relative density of the liquid **Solution:** $\tau = \mu \frac{du}{dy} \therefore \quad 0.6 = \mu \ \ x\,1.5 \ , \therefore \ \mu = \ 0.4\ Pa.s$ $\nu = \frac{\mu}{\rho}, \quad \therefore \rho = 0.4 / (4.65\ x\ 10\text{-}4) = 860.22\ kg/ m^3$ $\therefore$ Relative density $= \frac{\rho}{\rho_{water}} = \frac{860.22}{1000} = 0.86$

Ex.1.3	A Newtonian fluid of kinematic viscosity 2.528 stokes flows over a flat horizontal plate of surface area 0.8 m2. Velocity at y meters from plate is given as, $u = 2y — 2y^3$ m/s. If shear force acting on the plate is 0.352 N, find specific weight and specific gravity of liquid.

	Solution: As, $u = 2y - 2y^3$, Velocity gradient $= \frac{du}{dy} = 2 - 6y^2$ $\therefore$ At the start of plate ($y = 0$), $\frac{du}{dy} = 2$ $$\text{Also, } \tau = \frac{F}{A}, \quad F = \tau . A = \mu \frac{du}{dy} . A = \nu . \rho . \frac{du}{dy} . A$$ $\therefore 0.352 = (2.528 \times 10^{-4}) \times \rho \times 2 \times 0.8$ $\therefore \rho = 870.253 \text{ kg/ m}^3$ Specific Weight $= \gamma = \rho g = 870.253 \times 9.81 = 8531.18 \text{ N/ m}^3$ Specific Gravity $= S = \frac{8531.18}{9810} = 0.87$

Ex.1.4	A piston of 50 mm diameter moves within a cylinder of 50.1 mm bore. Determine percent decrease in the force necessary to move the piston when lubricating oil in the gap of piston and cylinder warms up from 40°C to 1100 ^{0}C. Use the following viscosity values.

Temperature ^{0} C.	40	60	80	100	120
Viscosity Pa-s	0.0053	0.0038	0.003	0.0029	0.0020

Solution:

Plot the graph of Viscosity versus Temperature as below.

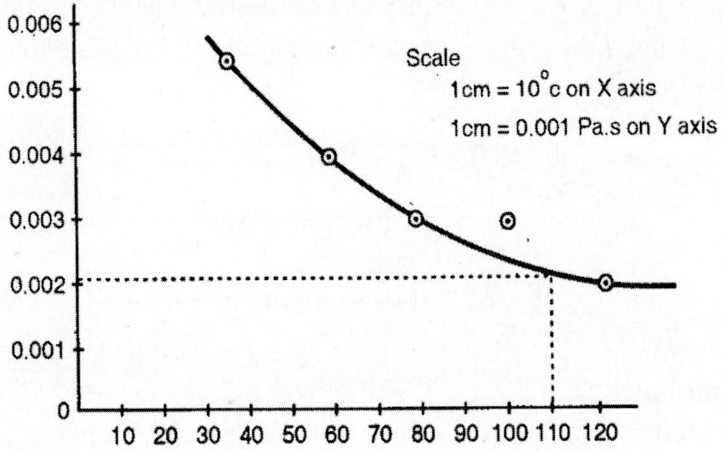

Fig. Ex. 1.4 Graph of Viscosity versus Temperature

From the above graph, the required values of viscosity can be achieved,

$$\therefore \mu \text{ at } 110^0\text{C} = 0.0023$$

Force required to move the piston at 400C = F1

$$\therefore F_1 = \mu \frac{du}{dy}.A = 0.0053.\frac{du}{dy}.A$$

Force required to move the piston at 1000C = F2

$$\therefore F_2 = \mu \frac{du}{dy}.A = 0.0023.\frac{du}{dy}.A$$

As , $(\frac{du}{dy}.A)$ remains same in both the above forces, percentage change

(decrease) in the force = $\frac{F_1 - F_2}{F_1}$ x 100 = $\frac{0.0053 - 0.0023}{0.0053}$ x 100 = 56.6%

Ex.1.5	There are two parallel plates 0.6 mm apart with the gap is filled by an oil of viscosity 1.5 Ns-m^2. The upper plate is moving with 3 m/s to the right while the

lower plate is moving with 3 m/s to the left. Determine the shear stress on both the plates if the velocity varies linearly from one plate to another.

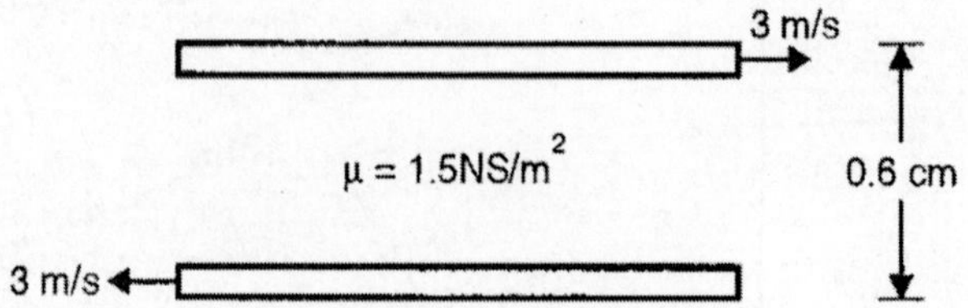

Fig. Ex. 1.5

Solution:

Velocity varies from “-3 m/s” (left) to “3 m/s” (right)

∴ Total variation in velocity is 6 m/s. Then as per Newton’s law of viscosity,

$$\tau = \mu \frac{du}{dy} = 1.5 \text{ x } 6 / (0.6 \text{ x } 10\text{-}3) = 15000 \text{ N/m}^2$$

Ex.1.6 A square metal plate of 1.5 m side and 1.5 mm thick weighs 50 N. It is to be lifted through a vertical gap of 25 mm of infinite extent. The oil in the gap has a specific gravity of 0.95 and viscosity of 2.5 pascal. If the metal plate is to be lifted at a constant speed of 0.1 m/s, determine the force and power required.

Solution:

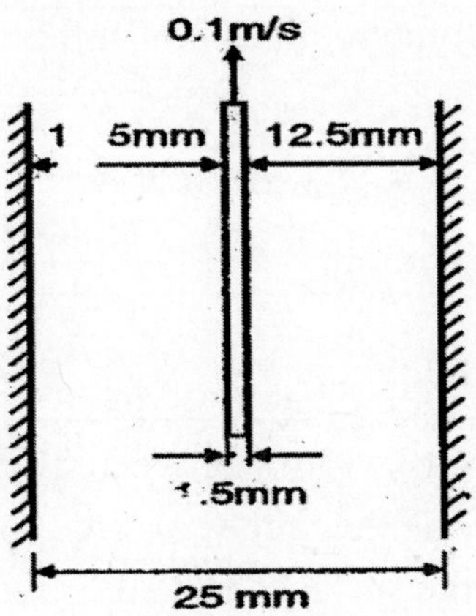

Fig. Ex. 1.6

Let us assume that the plate is placed centrally.

Submerged weight of Plate = WS = Weight of plate — Weight of liquid displaced by plate.

WS = 50 — (Volume of liquid displaced x specific weight of liquid)

= 50 – (1.5 x 1.5 x 0.0015 x 0.95 x 9810) = 18.55 N .

Shear force on left hand side of the plate = FL

$F_L = \tau . A = \mu \frac{du}{dy} . A = 2.5 \times 0.1 / ((12.5-0.75) \times 10^{-3}) \times 1.5 \times 1.5 =$ 47.87 N

By symmetry shear force on right hand side of the plate will be same as left hand side. $\therefore F_R = 47.87$ N

Therefore, Total force required to lift the plate

F = 18.55 + 47.87 + 47.87 =114.29 N.

Ex.1.7 A space of 2.5 cm width, between two large plane surfaces is filled with glycerine. Determine the force required to drag a very thin plate of surface area 0.75 m^2 between the surfaces at a speed of 0.5 m/s.

(i) If the plate remains equidistant from the two surfaces.

(ii) If it is at a distance of 1 cm from one of the surfaces.

Take dynamic viscosity = 0.705 Ns/ m^2.

Solution:

(i) If the plate remains equidistant from the two surfaces.

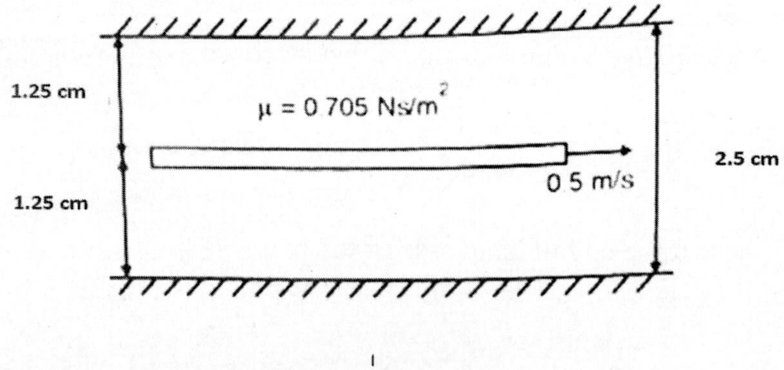

Fig. Ex. 1.7 (i)

$$\tau = \mu \frac{du}{dy} = 0.705 \text{ x} 0.5/(1.25 \text{ x } 10^{-2}) = 28.2 \text{ N/m}^2$$

$$F = \tau . A = 28.2 \text{ x } 0.75 = 21.15 \text{ N}$$

Total drag 1.25 cm 15 = 42.30 N

	(ii) If it is at a distance of 1 cm from one of the surfaces. 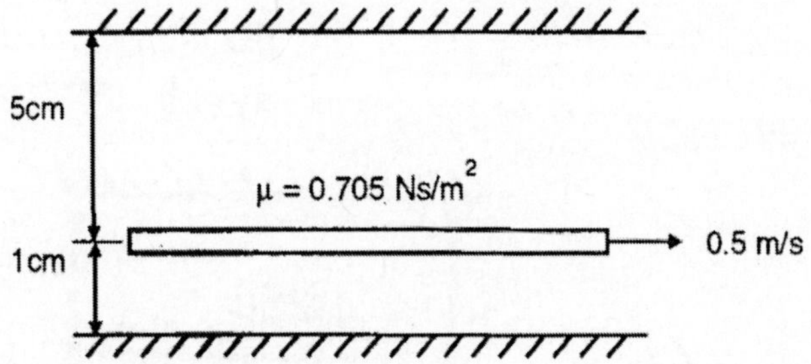 **Fig. Ex. 1.7 (ii)** $\tau_l = \mu \frac{du}{dy} = 0.705 \times 0.5 / (1 \times 10^{-2}) = 35.25 \text{ N/m}^2$ $F_l = \tau . A = 35.25 \times 0.75 = 26.44 \text{ N}$ $\tau_u = \mu \frac{du}{dy} = 0.705 \times 0.5 / (1.5 \times 10^{-2}) = 23.5 \frac{N}{m^2}$ $F_u = \tau . A = 23.5 \times 0.75 = 17.625 \text{ N}$ Total drag = Fl + Fu = 26.44 + 17.625 = 44.065

Ex.1.8	Through a narrow gap of large extent, a thin plate is pulled with constant speed. Liquids of velocities μ and 1.25 μ are filled in the gap below and above respectively. Assume the gap and plate are horizontal. Find distance of plate from lower surface of gap such that shear force either side of plate is the same, in terms of h. **Solution:** Let 'y' be the distance of the plate from lower surface and the plate moves with constant speed u

	Shear force on lower side of the plate, $F_1 = \tau . A$ $$F_1 = \mu \frac{du}{dy} . A = \mu \times \frac{u}{y} . A$$ 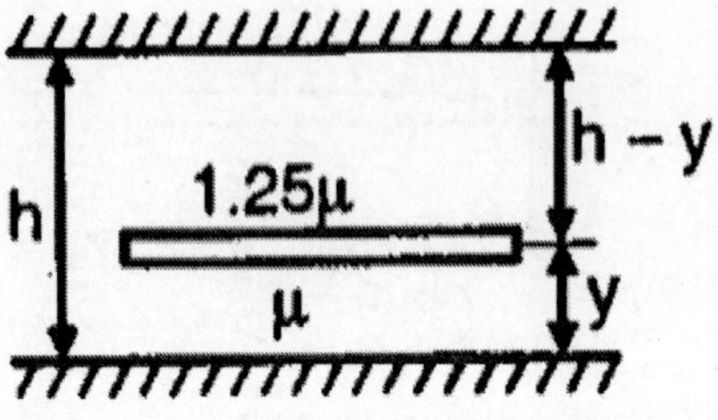 **Fig. Ex. 1.8** Shear force on lower side of the plate, $F_2 = \tau . A$ $$F_2 = \mu \frac{du}{dy} . A = 1.25\, \mu \times [\frac{u}{(h-y)}] . A$$ But , F1 =F_2 $$\therefore\ \mu \times \frac{u}{y} . A = 1.25\, \mu \times [\frac{u}{(h-y)}] . A$$ $$\therefore\ h - y = 1.25\, y$$ $$h = 2.25y$$ $$y = \frac{h}{2.25}\ m = 0.44\, h \quad \text{from lower surface}$$

Ex.1.9	Through a very narrow gap of thickness 'h' a thin flat plate of very large extent is being pulled at a constant velocity V. On one side of the plate lies oil of viscosity μ and on the other side is oil of viscosity kμ. Calculate the position of the plate so that the drag force on it is minimum.

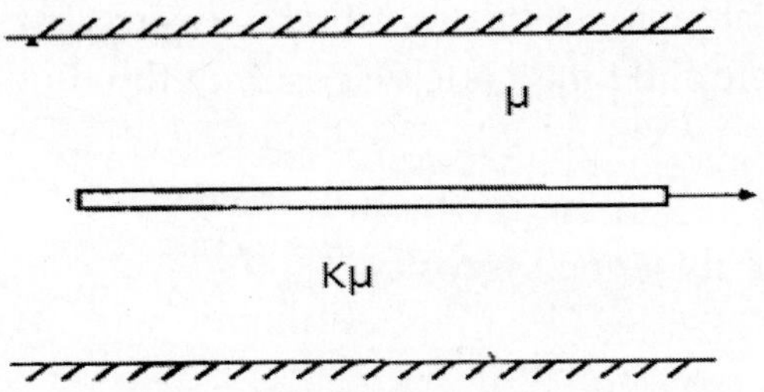

Fig. Ex. 1.9

Solution:

Let 'y' be the distance of thin flat plate from the top.

Shear stress on top of the plate = $\tau_1 = k\mu \text{ x} \frac{V}{y}$

Shear stress on top of the plate = $\tau_2 = \mu \text{ x} \frac{V}{h-y}$

Total Drag force acting F on the plate = $A\,(\tau_1 + \tau_2) = V\,A\,\mu\left(\frac{k}{y} + \frac{1}{h-y}\right)$

For minimum drag, $\frac{dF}{dy} = 0$

$$\therefore -\frac{k}{y^2} + \frac{1}{(h-y)^2} = 0$$

$$\therefore \frac{y^2}{(h-y)^2} = k, \therefore \frac{y}{h-y} = \sqrt{k},$$

$\therefore y = (h-y)\sqrt{k}$ -----this is the position of the plate from the top.

Ex.1.10	A rectangular plate of 2 m² area requires a force of 4 N to maintain a speed of 0.25 cm/sec while sliding over a fixed plate. The gap between the fixed and the

moving plate is 0.5 mm and filled with a fluid of specific gravity 0.8. Determine the dynamic and kinematic viscosity of the fluid in poise and stokes respectively.

Solution:

(1) For dynamic viscosity

Shear stress = $\tau = \frac{F}{A} = \frac{4}{2} = 2\ N/m^2$

Also, $\tau = \mu \frac{du}{dy}$, $\therefore\ 2 = \mu \frac{du}{dy} = \mu\,((0.25-0) \times 10^{-2}) / (0.5 \times 10^{-3})$

$\therefore \mu = 0.4\ Ns/m^2$

1 poise = $0.1\ Ns/m^2$, $\therefore\ \mu = 4$ poise

(2) For kinematic Viscosity, $\nu = \frac{\mu}{\rho}$

Specific gravity = Relative density = $\frac{\rho_{liquid}}{\rho_{water}}$

$\therefore \rho_{liquid} = 0.8 \times 1000 = 800$

$\therefore \nu = \frac{0.4}{800} = 5 \times 10^{-4}\ m^2/s$

1 stokes = $1 \times 10^{-4}\ m^2/s$, $\nu = 5$ stokes

Ex.1.11 A glass plate of 30 cm^2 rest on another plate with a film of oil 0.05 cm between them. A weight of half a kilogram is kept on the top plate. If the top plate starts sliding with a velocity of 1.5 cm/sec when both the plates are tilted at an angle 20° with the horizontal what is the viscosity of the liquid?

Solution:

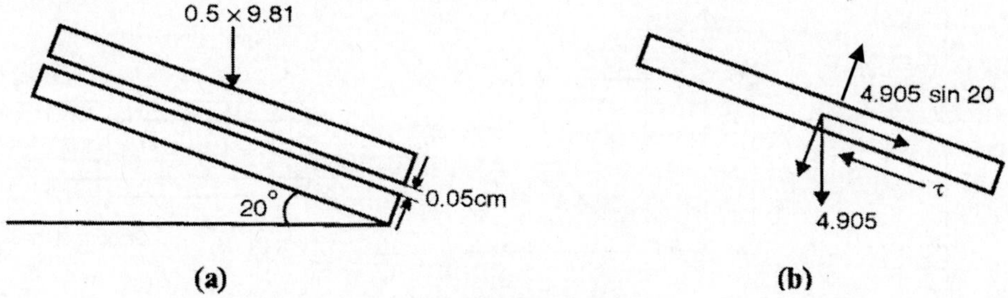

Fig. Ex. 1.11

The viscous shear force will be balanced by component of weight responsible for motion.

Therefore, Viscous shear force = Weight component

F = 0.5 x 9.81 x sin 20

$\therefore$ F = τ x A = 0.5 x 9.81 x sin 20 = 1.677 N

Also, $F = \mu \frac{du}{dy} A = 1.677$, $\therefore \ \mu \times \frac{1.5 \times 10^{-2}}{0.05 \times 10^{-2}} \times 30 \times 10^{-4} = 1.677$

$\therefore \mu = 18.64 \quad N\text{-}s/m^2$

x.1.12 A large plate moves with speed of V_O over a stationary plate on oil. If the velocity profile is that of a parabola, with the oil at the plates having the same velocity as that of the plates, determine the shear stress on the moving plate from the oil. If a linear profile is assumed, what will be the shear stress on the upper plate?

Solution:

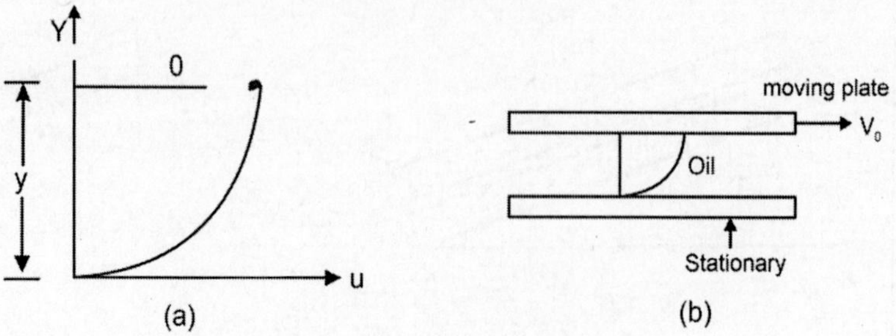

Fig. Ex. 1.12

Consider the parabolic velocity profile has vertex at the moving plate.

Then the velocity profile can be given by,

$$u = ay^2 + by + c, \quad \text{-----------------------}(1)$$

where a, b, and c are constants.

(a) at $y = 0$, $u = 0$

(b) at $y = y$, $u = V_0$ m/s

(c) at $y = y$, $\frac{du}{dy} = 0$

Then, substituting (a) in equation (1) , $c = 0$

substituting (b) in equation (1) , $V_0 = ay^2 + by$ ----------(2)

substituting (c) in equation (1) , $\frac{du}{dy} = 2ay + b = 0$ ------(3)

$$\therefore \text{shear stress} = \tau = \mu \frac{du}{dy} = \mu\,(2ay + b)$$

if the variation is linear,

$$\text{i.e } u = ay + b \; ----------- (4)$$

(d) at $y = 0$, $u = 0$

(e) at $y = y$, $u = V_0$ m/s

substituting in equation (4),

$b = 0$ and $u = ay$

$V_0 = ay$ and thus, $\frac{du}{dy} = a$

$\therefore$ shear stress $= \tau = \mu a$

Ex.1.13 A shaft 6.0 cm in diameter is being pushed axially through a bearing sleeve of diameter 6.02 cm and 40 cm long. The clearance, assumed uniform is filled with oil whose properties are kinematic viscosity$\nu = 0.003\ m^2 / s$ and specific gravity (g) = 0.88. Estimate the force required to pull the shaft at steady velocity of 0.4 m/s.

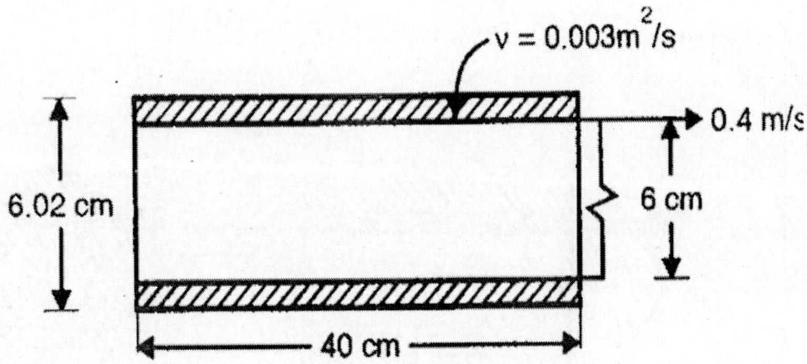

Fig. Ex. 1.13

Solution:

$$\text{Shear Stress} = \tau = \mu \frac{du}{dy}$$

$$\mu = \nu \times \rho = 0.003 \times 0.88 \times 1000 = 2.64\ Ns/m^2$$

$$\rho_{fluid} = g \times \rho_{water}$$

$$\tau = 2.64 \times \frac{0.4}{0.01 \times 10^{-2}} = 10560\ N/m^2$$

$$F = \tau \text{ x } A = 10560 \text{ x } \pi \text{ x } 0.06 \text{ x } 0.4$$

(shear force acts on the circumference of shaft πDL)

$$F = 796.21 \text{ N}$$

Ex.1.14 Lateral stability of a long shaft 150 mm in diameter is obtained by means of 250 mm stationary bearing having an internal diameter of 150.25 mm. If the space between the bearing and shaft is filled with lubricant having viscosity 0.245 N-s / m^2, what power will be required to overcome the viscous resistance when the shaft is rotated at 180 rpm.

Solution:

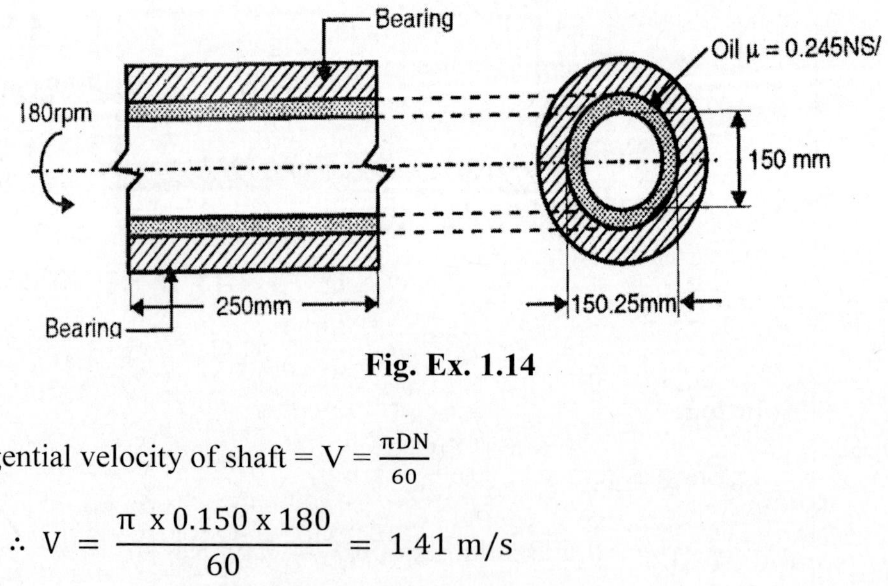

Fig. Ex. 1.14

Tangential velocity of shaft = $V = \frac{\pi DN}{60}$

$$\therefore V = \frac{\pi \text{ x } 0.150 \text{ x } 180}{60} = 1.41 \text{ m/s}$$

As, du = V-0 = 1.41 m/s and

dy = change in distance in vertical plane = $\frac{150.25-150}{2}$ = 0.125 mm

	$\therefore \tau = 0.245 \text{ x} \dfrac{1.41}{0.125 \text{ x } 10^{-3}} = 2763.6 \text{ N/ m}^2$ Shear force acting on the shaft F = 2763.6 x π x D x L = 2763.6 x π x 0.15 x 0.25 = 325.58 N $\text{Torque} = \text{T} = \text{F x } \frac{D}{2} = 325.58 \text{ x } \frac{0.15}{2} = 24.42 \text{ N}$ $\text{Power lost} = \dfrac{2\pi N T}{60} = \dfrac{2\pi \text{ x } 180 \text{ x } 24.42}{60} = 460.27 \text{ watt.}$

Ex.1.15	A flywheel of 50 kg mass, radius of gyration 20 cm is mounted at the middle of a shaft 3 cm in dimeter. The shaft is supported between two bearings, each 6 cm long. The clearance between the shaft and bearing of 0.05 mm is filled with an oil of viscosity (μ) 0.2 poise. Calculate the angular retardation of the shaft flywheel system due to frictional effects at a nominal speed of 1200 rpm. **Solution**: 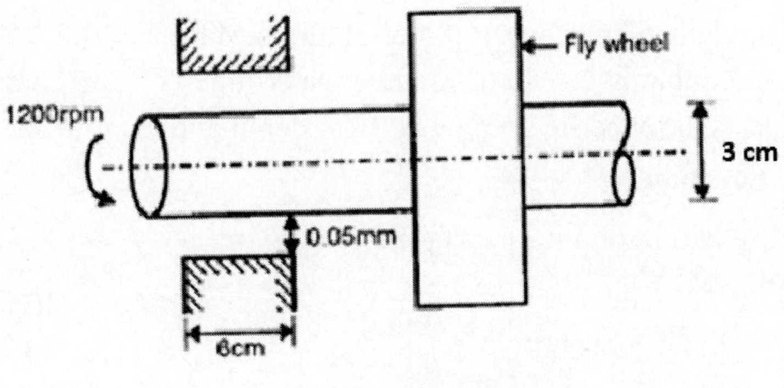 **Fig. Ex. 1.15**

	shear stress $= \tau = \mu \frac{du}{dy} = 0.2 \times 0.1 \times \frac{u}{y}$ $u = \frac{\pi DN}{60} = \frac{\pi \times 0.03 \times 1200}{60} = 1.885$ m/s $\tau = 0.2 \times 0.1 \times \frac{1.885}{0.05 \times 10^{-3}} = 753.98\ N/m^2$ shear force acting on shaft $F = \tau \times A = 753.98 \times \pi \times d \times L = 753.98 \times 3.14 \times 3 \times 10^{-2} x6 \times 10^{-2}$ F = 4.26 N Torque = T = F x r = 4.26 x 15 x 10-3 = 0.064 Nm But T = I x α (α is the retardation) And I = m k2 = 50 x (0.2)2 = 2 kg-m² ∴ 0.064 = 2 x α (assume α is in radian / sec²) ∴ α = 0.032 rad/ sec²

Ex.1.16	At a depth of 9 km in the ocean the pressure is 9.5×10^4 kN/m². The specific weight of the ocean water at the surface is 10.2 kN/m³ and its average bulk modulus is 2.4×10^6 kN/m². Determine (1) The change in specific volumeand (2) The specific volume at 9 km depth and (3) The specific weight at 9 km depth **Solution**: Specific volume at surface = $Vs = \frac{1}{\gamma} = \frac{1}{10.2 \times 10^3} = 9.804 \times 10^{-5}\ m^3/N$ $K = -\frac{dp}{\left(\frac{dV}{V}\right)}$ $2.4 \times 10^6 \times 10^3 = \frac{9.5 \times 10^4 \times 10^3}{dV} \times 9.804 \times 10^{-5}$ ∴ $dV = -3.81 \times 10^{-6}$ (decrease)

	Specific volume at 9 km depth = $9.804 \times 10^{-5} - 3.881 \times 10^{-6} = 9.416 \times 10^{-5}$ m^3 /N Specific weight at 9 km depth = $\frac{1}{9.416\text{x } 10^{-5}}$ = 10620.302 N/m^3

Ex.1.17	A pressure vessel has an internal volume of 0.5 m^3 at atmospheric pressure. It is desired to test the vessel at 3000 bar by pumping water into it. The estimated variation in the change of empty volume of the container due to pressurization to 3000 bar is 0.6 %. Calculate mass of water to be pumped into the vessel to attain the desired pressure level, given the bulk modulus of water as 2000 MPa. **Solution**: Change in volume inside the container due to addition of fluid = dV_1 $dV_1 = -\frac{VdP}{K} = \frac{0.5 \text{ x } (3000-1) \text{ x } 10^5}{2000 \text{ x } 10^6} = 0.075 \text{ m}^3$ (decrease) Change in volume of container due to change in dimensions = $dV_2 = \frac{0.6}{100} \text{ x } 0.5 = 0.003 \text{ m}^3$(decrease) Total volume of water accommodated = V = 0.075 + 0.003 = 0.078 m^3 Mass of water pumped in = 1000 x 0.078 = 78 kg .

Ex.1.18	If bulk modulus of water is 2.2 x 109 Pa, what pressure is required to reduce the volume of water by 6%. **Solution**:

$$K = -\frac{dp}{\left(\frac{dV}{V}\right)}, \quad \left(\frac{dV}{V}\right) = \frac{6}{100}$$

$$\therefore\ 2.2 \times 10^9 = \frac{dp}{\left(\frac{6}{100}\right)}, \quad \therefore\ dp = 132 \times 10^6\ Pa = 132\ MPa$$

Ex.1.19

Determine the diameter of a droplet of water in mm if the pressure inside is to be greater than that of outside by 130 N / m². Surface tension of water σ = 7.26 x 10-2 N/m.

Solution:

$$p = \frac{4\sigma}{d}, \qquad \therefore\ 130 = \frac{4 \times 7.26 \times 10^{-2}}{d}$$

$$\therefore\ \text{diameter of the droplet} = d = 2.23 \times 10^{-3}\ m = 2.23\ mm$$

Ex.1.20

Determine the pressure inside soap bubble of 25 mm diameter if the tension in the film is 0.5 N/m.

Solution:

$$p = \frac{8\sigma}{d} = \frac{8 \times 0.5}{25 \times 10^{-3}} = 160\ N/m^2$$

Ex.1.21

Calculate the capillarity rise or fall in a glass tube of 4 mm diameter when immersed in (a) water and(b) mercury.

The temperature of the liquid is 200^0C and the surface tension of water and mercury at 200C in contact with air is 0.0075 kg/m and 0.05 kg/m respectively. The contact angle for water and mercury may be taken as 00 and 1300respectively. Derive any equations used.

	Solution: $$h_{water} = \frac{4\sigma\cos\theta}{\gamma d} = \frac{4 \times 0.0075 \times 9.81 \times \cos 0}{9810 \times 4 \times 10^{-3}} = 7.5 \times 10^{-3}\ m$$ ∴ Capillary rise of water = 6.495 mm $$h_{mercury} = \frac{4 \times 0.05 \times 9.81 \times \cos 130}{13.6 \times 9810 \times 4 \times 10^{-3}} = -2.36 \times 10^{-3}\ m$$ ∴ Capillary rise of mercury = 2.36 mm

Ex.1.22	The diameters of two glass limbs of a differential U tube manometer were found to be 5 mm and 8 mm respectively. In an experiment the differential pressure readings of 50, 100, 250, 400 and 500 mm were indicated by the manometer. Determine the percentage error caused by the capillary effect. Surface tension of water σ = 0.0736 N/m and angle of contact θ=0° **Solution**: As, $$h_{water} = \frac{4\sigma\cos\theta}{\gamma d} = \frac{4\sigma}{\gamma d} ------ (\theta = 0 \quad \text{for water})$$ $$h_{water} = \frac{4 \times 0.0736}{9810 \times 0.005} = 6 \times 10^{-3}\ m = 6\ mm$$ Therefore, new readings will be 44, 94, 244, 394 and 494 respectively. ∴ percentage error, for 50 mm , $= \frac{100 \times 6}{50} = 12\ \%$ for 100 mm , = 6 % for 250 mm , = 2.4 % for 400 mm , = 1.5 % for 500 mm , = 1.2 %

Ex.1.23	A capillary tube of 0.5 mm bore stands vertically in a vessel containing a liquid of specific gravity 0.8 and surface tension 30 dynes/cm. The angle of contact of the liquid with the tube is zero. Find rise of liquid in the tube. **Solution:** σ = 30 dyne / cm = $30 \times 10^{-5} \times 100 = 30 \times 10^{-3}$ N/m Capillary rise = h = $\frac{4\sigma\cos\theta}{\gamma d} = \frac{4 \times 30 \times \cos 0 \times 10^{-3}}{0.8 \times 9810 \times 0.5 \times 10^{-3}}$ = 0.03058 m = 30.58 mm

UNIT SUMMARY

The unit gave information of properties of fluids which is essential to understand the future topics in fluid mechanics. The difference between liquids and gases, classification of fluids and units of measurements were also discussed. As discussed in the "Introduction" the next unit is devoted to fluid statics.

1.6 Exercise

1.6.1 Objective Questions:

1. Which among the following is a Newtonian fluid?
   (a) Slurry (*b*) paste (*c*)gel (*d*) benzene
   Ans: d

2. Mass of 6 m^3 of water at 20^0 C and 1 atm., pressure is 6000 Kg and viscosity under similar condition is 1 centipoise then kinematic viscosity of water is
   a) 10^{-3} m^2/s b) 10^{-4} m^2/s c) 10^{-5} m^2/s d) 10^{-6} m^2/s
   Ans: d

3. Volume of sample of water is reduced by 1 percent when pressure is increased by 22 MPa. The bulk modulus of elasticity of the sample in MPa is
   a) 2.2 b) 220 c) 2200 d) 0.22
   Ans: c

4. Newton's law of viscosity

(a) Defines the fluid property called viscosity.

(*b*) Can be applied only if velocity profile is linear

(*c*) Applies only at the interface between a solid and fluid

(*d*) Is inconsistent with the no-slip condition

Ans: a

5. The gauge pressure inside 2 mm diameter raindrop, taking surface tension of water -air interface as 0.07 N/ m is

a) 0.14 N/m^2 b) 140 N/m^2 c) 70 N/m^2 d) 280 N/m^2

Ans: b

6. The gauge pressure inside a droplet of water of certain diameter is 70 N/m^2. The gauge pressure inside a droplet of twice the diameter under similar conditions is

a) 35 N/m^2 b) 140 N/m^2 c) 17.5 N/m^2 d) 280 N/m^2

Ans: a

7. Toothpaste is

(a) Bingham Plastic

(b) Pseudo Plastic

(c) Newtonian Fluid

(d) Dilatants

Ans: a

8. The capillary rise in 1.5 mm tube immersed in liquid is 12mm. The capillary rise in 2 mm diameter tube immersed in the same liquid will be

a) 9 mm b) 16 mm c) 20 mm d) 24 mm

Ans: a

9. The viscosity of most fluids

(a) Decreases with decrease in temperature.

(b) Increases with decrease in temperature

(c) Decreases with increase in temperature

(d) Increases with increase in temperature.

Ans : c

10. The mass density of water on earth ($g = 9.81 \text{ m/ s}^2$) is 1000 Kg/ m^3. Its mass density on the moon where the gravitational acceleration is 1/6th that of the earth , will be
a) 166.67 Kg / m^3 b) 1000 Kg / m^3 c) 6000 Kg / m^3 d) none of these
Ans: b

11. The mass of an object is 10 kg. The gravitational acceleration at a location is 5 m/s^2. The specific weight is
(a) 2 N (*b*) 15 N (*c*) 5 N (*d*) 50 N
Ans: d

12. For an ideal fluid flow Reynolds number is
(a) Infinite,
(b) Zero
(c) 2100
(d) One

Ans : a

13. The dynamic viscosity is 1.2×10^{-4} Ns/m^2. The density is 600 kg/m^3. The kinematic viscosity in m^2/s is
(a) 72×10^{-3} (*b*) 20×10^{-8} (*c*) 7.2×10^3 (*d*) 70×10^6
Ans: b

14. The velocity gradient is 1000/s. The viscosity is 1.2 × 10–4 Ns/m^2. The shear stress is
(a) 1.2 × 10–1 N/m (*b*) 1.2 × 10–7 N/m^2 (*c*) 1.2 × 102 N/m^2 (*d*) 1.2 × 10–10 N/m^2

Ans: a

15. If the specific weight of water is taken as 9.81 KN/m^3 and the specific gravity of mercury is 13.56, then the density of water will be

(a)1000kg/m^3 (*b*) 1080 kg/m^3 (*c*) 981 kg/m^3 (*d*) 9810 kg/m^3

Ans: a

16. The effect of Cavitations is due to:
(a) High velocity
(b) Low barometric pressure
(c) High pressure
(d) Low pressure

Ans : c

1.6.2 Theory Questions:

1. Define fluid. Distinguish between ideal fluid and real fluid? Give one example of each.
2. Define following properties with their SI units: Specific weight, mass density, specific gravity, specific volume, bulk modulus, Capillarity, surface tension
3. Explain why the following statements are right or wrong with detail reasoning .
   (a) Ideal fluid can sustain a shearing stress when in motion.
   (b) Fluids cannot sustain shearing stress when at rest.
   (c)
4. Draw the stress strain curve for the following fluids and discuss the behaviour of each fluid under external shear force
   (a) Newtonian fluid,
   (b) Pseudoplastic fluid,
   (c) Dilatant fluid,
   (d) Bingham fluid,
   (e) Plastic,
   (f) Non-Newtonian fluid.
5. Explain the difference between dynamic viscosity and kinematic viscosity of a fluid.
6. Explain the term vapour pressure and discuss its relation with the cavitation.
7. Define capillarity and prove that It is given by $h = \frac{4\sigma}{\gamma_l\, d}$

8. Explain why viscosity of liquids decreases with rise in temperature while gases increases with rise in temperature.
9. What is kinematic viscosity? Why is it so called? Give it's units and dimensions,
10. State and explain Newton's law of viscosity.
11. What is cohesion and adhesion in fluids?
12. What is capillarity? What is it due to? Derive an expression for the capillary rise
13. Derive an expression between pressure 'P' inside a free jet of liquid and surface tension 'σ'.
14. Derive expression for pressure P' inside the soap bubble, droplet and surface tension
15. 'σ'.
16. Distinguish between gases and liquids.
17. Distinguish between Newtonian and Non-Newtonian fluids.
18. Distinguish between Kinematic and dynamic viscosity.
19. Explain compressibility in fluid flow.
20. Discuss why water shows capillary rise and mercury shows capillary depression
21. Explain the terms: Vapour pressure, Bulk Modulus
22. Explain the two applications of the following properties -one advantageous and other disadvantageous: Vapour Pressure, surface tension, Capillarity

1.6.3 Problems:

1. Three liters of liquid 23.7 N. Calculate its mass density, specific wight, specific gravity.
   (Ans: 805 kg/m^3, 7.9 KN/m^3, 1.242 x 10-3 m^3/Kg)

2. A plate 0.05mm distant apart from a fixed plate moves at 1.2m/s and requires a shear stress of 2.2 N/m^2 to maintain its viscosity. Find the viscosity of the fluid between the plates.
   (Ans: 9.16 x 10-5 N-s/ m^2)

3. An oil of kinematic viscosity having 1.25 x10-4 m^2/s and a specific gravity of 0.80. What is its dynamic viscosity in Kg/m-sec?
(Ans: 0.10 Kg/m-sec)

4. The space between two parallel pates kept 3 mm apart is filled with an oil of dynamic viscosity 0.2 poise. Determine the shear tresses on the fixed plate, if the upper one is moving with a velocity of 90 m/min.
(Ans: 10 N/m^2)

5. What is the pressure within a 1 mm diameter spherical droplet of water relative to it's outside atmospheric pressure? Assume surface tension for pure water to be 0.073 N/m.
(Ans: 292 N/m^2)

6. A 20 mm diameter soap bubble has an internal pressure of 27.576 N/ m^2greater than the outside atmospheric pressure. Determine the surface tension of that soap air double is in (N/m).
(Ans:0.0689 N/m)

7. A small circular jet of water of 2 mm diameter issues from an opening. What is the Pressure difference between inside and outside of the jet?
(Ans: 73.5 N/m^2)

8. In the Fig., if the fluid is glycerine at 20°C and the width between plates is 6 mm, what shear stress (in Pa) is required to move the upper plate at Velocity of 5.5 m/s? Note that glycerine viscosity =1.5 N- s/m^2.
(Ans: 1380 Pa)

9. The specific weight of water at ordinary pressure and temperature is 9.81 kN/m^3. The Specific gravity of mercury is 13.56. Compute the density of water and the specific weight and density of mercury.
(Ans:1000kg/m^3 ,133.0 kN/m^3, 1356 kg/m^3)

10. A space 16 mm wide between two large plane surfaces is filled with SAE 30 Western lubricating oil at 35°C . What force F is required to drag a very thin plate of 0.4 m^2 area

11. Between the surfaces at a speed v 0.25 m/s . (a) if the plate is equally spaced between The two surfaces, and (b) if t 5 mm?
(Ans:7.25 N, 8.436 N)

PRACTICAL: REDWOOD VISCOMETER

Objective:

1) To determine kinematic viscosity of a given liquid at a given temperature.

2) To study variation of kinematic viscosity with temperature.

Apparatus:

Redwood viscometer, oil (whose viscosity is to be measured), water, stopwatch, Thermometer

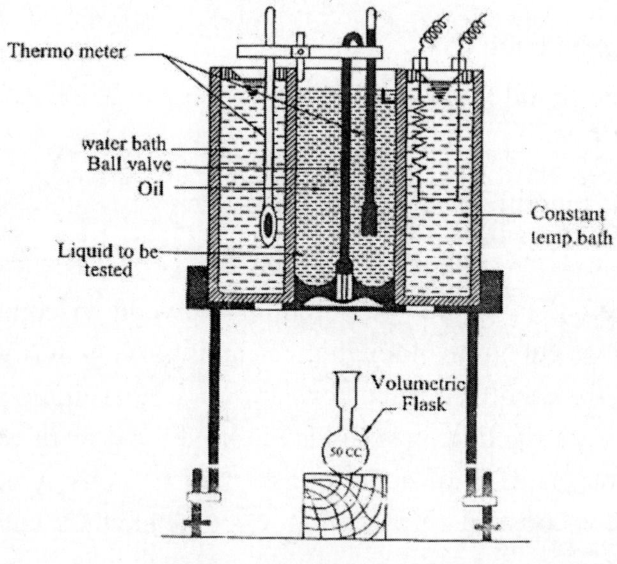

Figure: Red Wood Viscometer

Theory:

Viscosity is the property of fluid by virtue of which it offers resistance to the movement of one layer of fluid over the adjacent layer. Measurement of viscosity of Newtonian fluid can be done by using either Newton's law of viscosity or Hagen - Poiseuille equation or by Stokes's law. Redwood viscometer is an instrument which works on the basis of Hagen-Poiseuille's equation. It consists of a vertical cylinder provided with a small pipe orifice a at the centre of its base. The cylinder is surrounded by a water bath which can be electrically heated. The cylinder is filled with the liquid, the viscosity of which is to be determined and time required to pass 50 cc of that liquid at desired temperature is measured which in turn is used to calculate the viscosity.

Hagen-Poiseuille equation for a steady laminar through circular pipe can be written as

$$Q = (\Delta_p \pi D^4)/128\ \mu L \qquad \text{-------------------------(1)}$$

Put, Q=Discharge = Volume /time = V / t

D=Diameter of tube

Δ_p= pressure drop in tube =ρgh

μ=dynamic viscosity

ρ= density of fluid

g = gravitational acceleration

h= head under which liquid flows through a tube

ν=kinematic viscosity

Hence equation (1) is modified as

$$\mu/\rho = \nu = [(\pi\ hgD^4)/(128\ LV)] \text{ ---------------------(2)}$$

All quantities in equation (2) can be measured in redwood viscometer to determine kinematic viscosity. In equation 2 though the head is varying during test, its variation is over same range for each test, since constant volume of liquid is allowed to flow for each test. Let ' t' be the time in seconds required for flow of constant volume of liquid then all terms on RHS of equation (2) may be considered as constant and grouped to provide a constant to particular viscometer, then equation (2) can be modified as

$$\nu = c_1 t \qquad (3)$$

Equation (3) shows that kinematic viscosity varies directly with time. As the capillary rise is quite short, steady laminar flow condition will usually not exist in the capillary pipe provided in the viscometer. Thus, a correction factor will have to be incorporated in equation (3) in order to compensate terms of Hagen-Poiseuille equation in the analysis to obtain the correct value of ν of the liquid. The correction is (C_2/t). Then, equation (3) becomes,

$$\nu = C_1 t - (C_2 / t) \quad (4)$$

where t = time in seconds required to pass 50 cc of oil

Procedure:

1) Instrument is levelled with the help of foot screws and leveling tube in order to ensure uniform head of liquid over orifice
2) Close the orifice opening
3) The water bath is filled
4) Liquid (oil) whose viscosity is to be measured is filled up to the index mark
5) The steady temperature in liquid and water bath is measured
6) Orifice valve is opened when temperature of liquid and water bath is same and time required to collect 50cc of oil is noted
7) Above procedure is repeated for different temperatures by heating water bath

For I.S.I marked viscometers (values can be different for different equipment)

$C_1 = 0.0026$

$C_2 = 1.175$

Observations:

Sr. No.	Temperature in ^{0}C (T)	Time in seconds (t)

Sample calculation: For observation No.:

Kinematic viscosity in m^2/s, $\nu = C_1 t - (C_2/t)$ =

Tabulated Calculations:

Sr. No.	Temperature in ^0C (T)	Kinematic Viscosity (ν)in m^2/s

Graph: Plot the graph between 'Kinematic viscosity' on Y axis and 'temperature' on X axis and study the variation in Kinematic viscosity with respect to change in temperature.

Conclusion: Kinematic viscosity of a given liquid at room temp (0^c) is () m^2/s. Kinematic viscosity of a given liquid decreases with increase in temperature of fluid.

REFERENCES AND SUGGESTED READINGS

1. Hydraulics & Fluid Mechanics by Modi and Seth, Standard Book House
2. Theory and Applications of Fluid Mechanics—K. Subramanya- Tata McGraw
3. Fluid Mechanics by R.J. Garde, A.J Mirajgaonkar, SCITECH Publication
4. Fluid Mechanics by Streeter & Wylie, Tata McGraw Hill.
5. https://archive.nptel.ac.in/noc/courses/noc19/SEM2/noc19-ce28.

2 FLUID STATICS

UNIT SPECIFICS

Through this unit we have discussed the following aspects:

- *Introduction of fluid statics, pressure at a point, variation of pressure in fluid statics.*
- *Measurement of pressure, various pressure measuring devices.*
- *Hydrostatic forces on surfaces*
- *Total pressure on horizontal, vertical, inclined, curved surfaces.*

This is followed by large number of solved examples. The students are encouraged to solve the objective questions, long answer questions and numerical problems to judge ones understanding. A demonstration kind of practical (study experiment) on pressure measuring devices is included at the end followed by a list of references for additional reading.

RATIONALE

This unit introduces fluid statics considering basic concept of pressure, variation of pressure in fluid statics. To understood measurement of pressure in fluid statics, one must know various pressure measuring devices and the process of measurement of pressure using these devices and hence introduction of these devices is included here in this unit. Knowing the concept of pressure and its measurement, hydrostatic forces on surfaces, total pressure on horizontal as well as vertical, inclined and curved surfaces can be determined.

PRE-REQUISITES

Mathematics: Derivatives (Class XII)

Physics: Mechanics (Class XII)

Properties of fluids

UNIT OUTCOMES

List of outcomes of this unit is as follows:

(At the end of this unit, students will understand...)

U2-O1: Fluid statics, pressure at a point, variation of pressure in fluid statics.

U2-O2: Measurement of pressure, various pressure measuring devices

U2-O3: Hydrostatic forces on surfaces

U2-O4: Total pressure on horizontal, vertical, inclined, curved surfaces.

Unit-2 Outcomes	EXPECTED MAPPING WITH COURSE OUTCOMES (1-WeakCorrelation;2-Mediumcorrelation;3-StrongCorrelation)					
	CO-1	CO-2	CO-3	CO-4	CO-5	CO-6
U2-O1	2	2	-	-	-	-
U2-O2	2		-	-	-	-
U2-O3	2		-	-	-	-
U2-O4	2		-	-	-	-

2.1 Introduction:

Fluid static deals with study of action of forces on fluids at rest. When fluids are at rest, there is no relative motion between adjacent layers of fluid and thus no shear stress. The only forces which act on fluid at rest are the external pressure forces and the self-weight.

This chapter describes in detail fluid pressure, its measurement and hydro-static forces due to pressure acting on plain and curved surfaces

2.2 Pressure at the Point:

If an infinitesimally small area is considered inside a large fluid mass, the surrounding fluid exerts a force on that area. This force will always be normal to that area if the fluid is in static condition. The limiting ratio of this force with area is termed as 'Intensity of pressure' i.e.

$$P = \lim_{dA \to 0} \frac{dF}{dA} \quad (2.1)$$

However, the pressure is assumed to act uniformly all over the area. (This is similar to the concept of uniform normal stress distribution on solids)

$$\therefore \quad P = \frac{F}{A} \quad (2.2)$$

Where F is total force acting on Area A. It can also be seen that pressure is inversely proportional to the area. It means more pressure will act on less area keeping the force constant

Unit of pressure is Pascal (Pa) or kilo Pascal (kPa) or bar.

$$1 \text{ Pa} = 1 \text{ N/m}^2$$

$$1 \text{ kPa} = 103 \text{ }^{N}/\text{m}^2$$

$$1 \text{ bar} = 10^5 \text{ N/m}^2$$

2.3 Pascal's Law:

In case of stationary fluid in equilibrium, the pressure at a point acts equally in all the directions. This is Pascal's law of pressure.

Consider an infinitesimal fluid particle of dimensions δx, δy, δz along X, Y, Z directions respectively with δs as the length of slopping side (Refer figure 2.1).

Let P_x, P_y be the intensities of pressure along X and Y directions and P_s be the intensity normal to the slopping side. The free body diagram is shown in Fig, 2.1. As the element is in equilibrium, sum of forces in all directions must be zero.

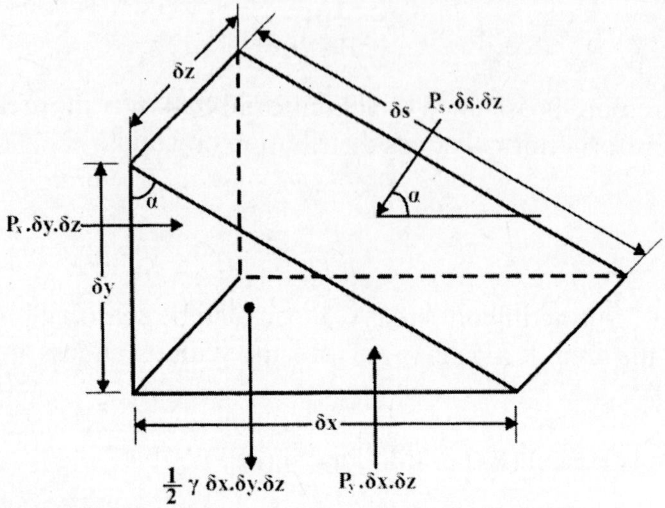

Fig. 2.1 Free body diagram of fluid element at rest

∴ $$\sum Fx = 0$$

∴ $$Px\ \partial y\ \partial z - P_s \cos\alpha\ \delta s\ \delta z = 0$$

But $$\delta s \cos\alpha = \delta y$$

∴ $$P_x\ \delta y\ \delta z = 0$$

∴ $$P_x = P_s \tag{2.3}$$

Similarly, $$\sum F_y = 0$$

∴ $$P_y\ \delta x\ \delta z - P_s \sin\alpha\ \delta s\ \delta z - \frac{1}{2}\gamma\ \delta x\ \delta y\ \delta z = 0$$

where $\frac{1}{2}\gamma\ \delta x\ \delta y\ \delta z$ is the weight of fluid element which will be neglected being of higher order than the other 2 terms. Also, $\delta s \sin\alpha = \delta x$

∴ $$P_y\ x\ \delta z = P_s\ \delta x\ \delta z = 0$$

∴ $$P_y = P_s \tag{2.4}$$

From Equations (2.3) and (2.4),

$$P_x = P_y = P_s$$

This proves that intensity of pressure, at a point, is equal in all directions. This principle is used to develop large forces by application of very small forces in equipment like hydraulic ram, hydraulic lift, etc.

2.4 Variation of Pressure in Static Fluid:

As mentioned earlier the only forces acting on fluid at rest are pressure force and gravity force. The gravity force expressed as weight of the fluid is constant and acts through the centre of gravity.

The pressure is same at a point in all directions as per Pascal's law. The variation of pressure with respect of height or depth of fluid mass can be found out using principles of static equilibrium.

Consider as elementary mass of static fluid in the form of a parallelepiped of dimensions δx, δy, δz as shown in Fig. 2.2. The forces acting on fluid mass are the weight of fluid element and pressure forces on all the six faces. Let 'p' be the pressure at the centroid 'o' of the element. The intensity of pressure of left face of the element be $\left[p - \left(\frac{\delta p}{\delta x}\right).\frac{\delta x}{2}\right]$ and on the right face be $\left[p + \left(\frac{\delta p}{\delta x}\right).\frac{\delta x}{2}\right]$.

$\Big(\frac{\delta p}{\delta x}$ indicates variation of pressure in 'x' direction, which is a for a length of $\frac{\delta x}{2}\Big)$

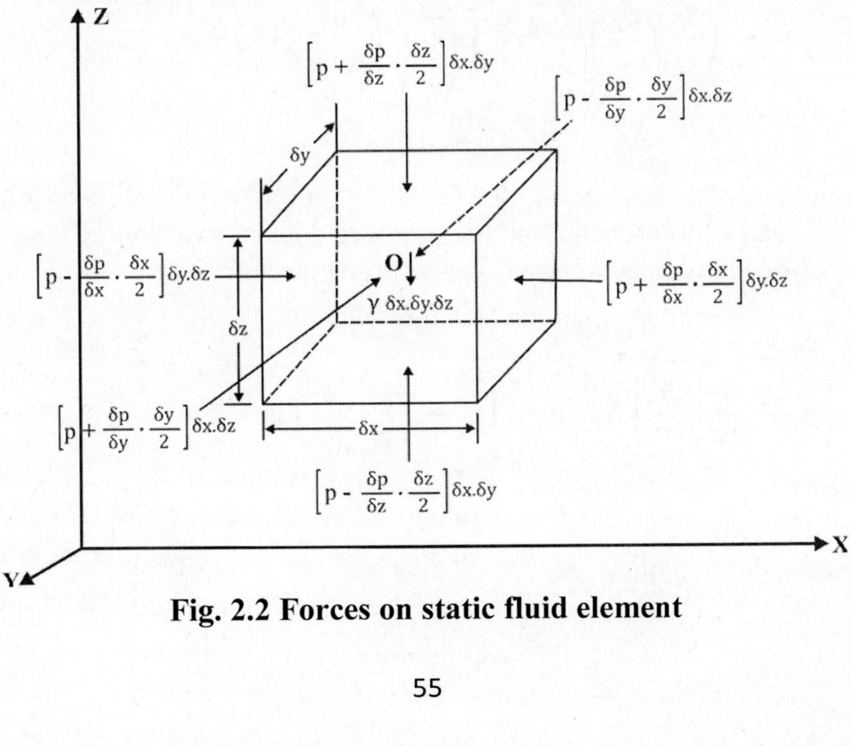

Fig. 2.2 Forces on static fluid element

$\therefore$ The force acting on left face = $\left[P - \left(\frac{\partial P}{\partial X}\right) . \frac{\partial X}{2}\right] . \delta y\, \delta z$ and

the force acting on right face = $\left[p + \left(\frac{\partial p}{\partial x}\right) . \frac{\partial x}{2}\right] \delta y\, \delta z$

Similarly forces acting on remaining 4 faces are calculated and shown in Fig. 2.2. Using equations of static equilibrium.

$$\sum F_x = 0$$

$$\left[p - \left(\frac{\partial p}{\partial x}\right)\frac{\partial x}{2}\right]\delta y\, \delta z - \left[p + \left(\frac{\partial p}{\partial x}\right)\frac{\partial x}{2}\right] \delta y\, \delta z = 0$$

$\therefore$
$$\left(\frac{\partial p}{\partial x}\right) \delta x\, \delta y\, \delta z = 0$$

$\therefore$
$$\frac{\partial p}{\partial x} = 0 \text{ (as } \delta x\, \delta y\, \delta z \neq 0) \qquad (2.5)$$

Similarly in 'y' direction

$$\sum F_y = 0$$

$$\left[p - \left(\frac{\partial p}{\partial y}\right)\frac{\partial y}{2}\right]\delta x\, \delta z - \left[p + \left(\frac{\partial p}{\partial y}\right)\frac{\partial y}{2}\right] \delta x\, \delta z = 0$$

$\therefore$
$$\frac{\partial p}{\partial y} = 0 \qquad (2.6)$$

Equations (2.5) and (2.6) indicate that the pressure does not vary in 'x' and 'y' direction. This confirms the already proved Pascal's law (section 2.3).

In 'z' direction.

$$\left[p - \frac{\partial p}{\partial z} . \frac{\partial z}{2}\right] \delta x . \delta y - \left[p + \frac{\partial p}{\partial z} . \frac{\partial z}{2}\right] \delta x . \delta y - \gamma\, \delta x\, \delta y\, \delta z = 0$$

$\therefore$
$$\frac{\partial p}{\partial z} = -\gamma$$

i.e. $\frac{\partial p}{\partial z} = -\gamma$ (as $\frac{\partial p}{\partial x} = \frac{\partial p}{\partial y} = 0$. p ⇒ f(z) only (2.7)

Equation (2.6) proves that pressure varies in vertical direction. For incompressible fluids, specific weight 'γ' is constant. Therefore Equation (2.7) can be integrated between two points.

$$\int_{p1}^{p2} dp = -\gamma \int_{z1}^{z2} dz$$

∴ $$p_2 - p_1 = -\gamma\,(z_2 - z_1)$$

∴ $$\frac{p2}{\gamma} + z_2 = \frac{p1}{\gamma} + z_1 \qquad (2.8)$$

The term $\frac{P}{\gamma} + z$ is known as the piezometric head. Thus for incompressible fluids piezometric head is constant.

Consider free surface of liquid in figure 2.3 along which pressure is constant and equal to the atmospheric pressure P_a. The free surface is at 'H' above the datum. At a point 'p' located 'h' below free surface or 'z' above datum (i.e. z + h = H) integrating Equation (2.8)

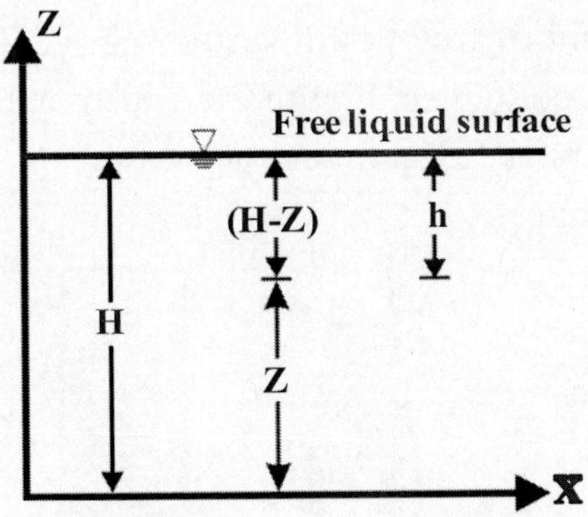

Fig. 2.3 Pressure at a point

$$p = -\gamma z + c$$

at $z = H,\ p = p_a$

∴ $$p_a = \gamma H + c$$

∴ $$c = p_a - \gamma H$$

∴ $$p = -\gamma z + p_a - \gamma H$$

∴ $$p = p_a + \gamma (H - z)$$

∴ $$p = p_a + \gamma h$$

In most of the problems pressure above atmospheric pressure is required.

∴ $$p = \gamma h \tag{2.9}$$

Equation (2.9) is known as 'Law of Hydrostatics'. Equation (2.9) can also be written as,

$$h = \frac{p}{\gamma}$$

Thus pressure can also expressed as equivalent pressure head 'h' of the liquid. In simple terms it can be said that due to pressure 'p' at a point, there is a rise 'h' of liquid or the rise 'h' of liquid is balancing the pressure 'p'. This concept of equivalent pressure head will be made explained more in the section on Piezometers and manometers.

Equation (3.8) also indicates that pressure varies linearly with the depth of flow below liquid surface or height of free liquid surface above the point and is independent of shape and size of the container. Figure 2.4 illustrates this point, wherein the rise of liquid in all containers is 'h' showing a pressure p = γh at the centre of the pipe irrespective of their shapes

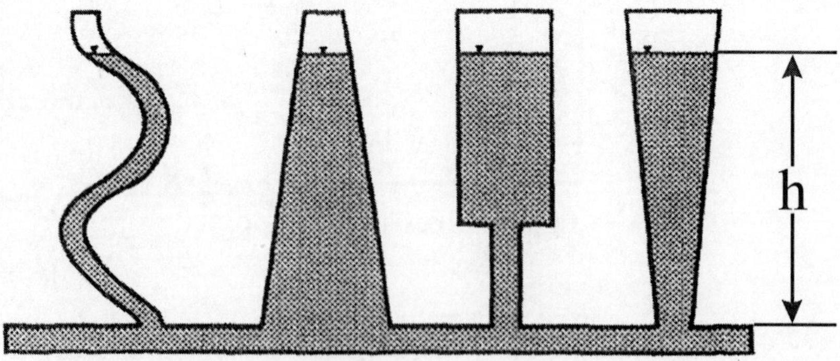

Fig. 2.4 Pressure independent of size and shape

Consider a column of water 'h_w' indicating pressure $p = \gamma_w h_w$

If the same pressure is measured with kerosene of specific gravity 0.8, the rise of kerosene column will be more than water. On the other hand, if an oil of specific gravity 1.6 is used to measure the same pressure, the rise of oil column will be less than that of water. This can explained as follow:

Refer figure 2.5 with rise of water column hw = 1 m.

As pressure is same.

$$P = \gamma_w h_w = \gamma_k h_k = \gamma_o h_o$$

$$h_w = \frac{\gamma_k}{\gamma_w} h_k = \frac{\gamma_o}{\gamma_w} h_o$$ where 'k stans for kerosene and 'o' stands for oil.

$$h_w = S_k h_k = S_o h_o \left(Specific\ gravity = \frac{\gamma_{fluid}}{\gamma_{water}}\right)$$

$$1 = 0.8\ h_k = 1.6\ h_o$$

$h_k = 1.25$ m and ho = 0.625 'm'

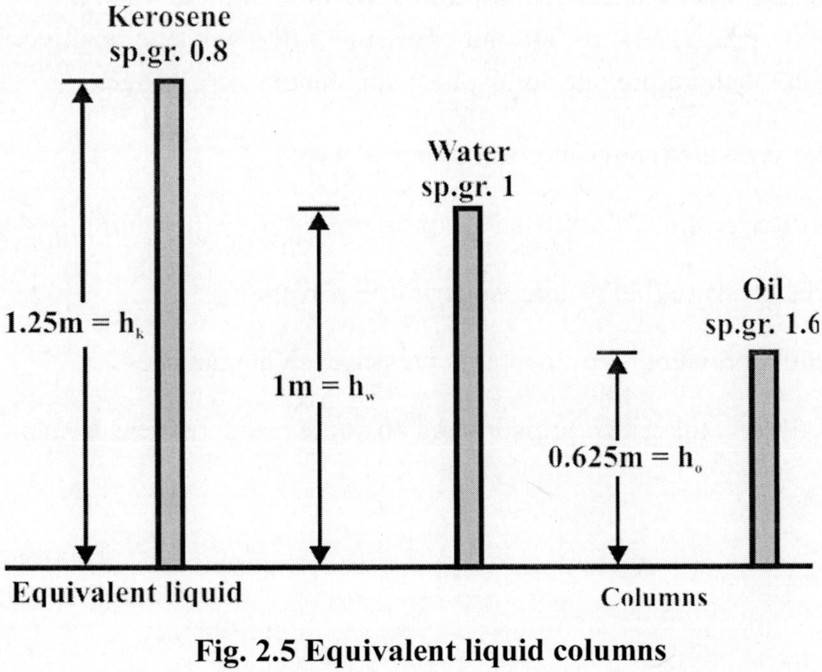

Fig. 2.5 Equivalent liquid columns

Thus lighter liquid will show higher rise than heavier liquids.

In general

$$h_{water} = S_{fluid} \times h_{fluid} \tag{2.10}$$

This concept of equivalent liquid column is very useful in pressure measuring devices like manometers. Equation (2.10) helps to express pressures measured by any fluid in terms of a common fluid say 'water'.

2.5 Measurement of Pressure:

Pressure is measured with respect to some datum. The datum can be local atmospheric pressure or absolute zero pressure (complete vacuum). The atmospheric pressure considered as a reference is 760 mm of Mercury i.e., 10336 mm of water or 101.33 kN/m^2. When the fluid pressure is measured with respect to atmospheric pressure as datum then it is known as gauge pressure. If the pressure is above atmospheric pressure, then it is called positive pressure. If the pressure is below atmospheric pressure, then it is called negative pressure or vacuum pressure. The pressure measured with respect to absolute zero or complete vacuum as datum is called absolute pressure. Thus, absolute pressure will always be positive. The atmospheric pressure is therefore one atmosphere absolute or zero gauge.

For positive pressure (above atmospheric pressure).

Absolute pressure = Atmospheric pressure + Gauge pressure

For negative pressure (below atmospheric pressure).

Absolute pressure = Atmospheric pressure – Vacuum pressure

Figure 2.6 depicts the gauge pressure and absolute pressure measurement.

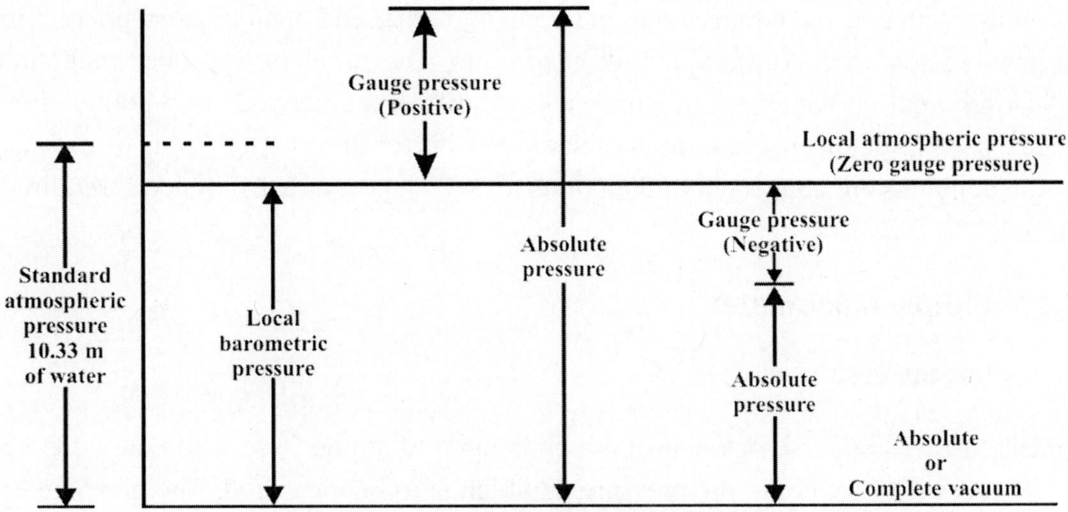

Fig. 2.6 Absolute and gauge pressure

2.6 Pressure Measuring Devices:

Pressure measuring devices are of two types:
1. Manometer 2. Mechanical gauge.

In this era of advanced equipment pressure transducers are also used to measure pressure very accurately. However, the scope of present article is limited to first two types only.

2.6.1 Manometer:

Manometers measures the fluid pressure by balancing it against a column of liquid in static equilibrium. These are generally glass tubes or tubes of any transparent material of diameter more than 6 mm, to avoid capillary effect as explained in unit 1. The liquid used to balance the fluid pressure is called as Manometric fluid. The manometric fluid should be immiscible with the fluid of which pressure is measured. Also the manometric fluid should have low vapour pressure, otherwise it will evaporate very quickly. Standard manometric liquids therefore are mercury, water, air, carbon-tetra-chloride. The choice of manometric liquid depends upon the magnitude of pressure, type of manometer, desired accuracy. Manometers are classified into i) Simple manometers ii) Differential manometers. Simple manometers measure pressure at a point, with one end connected to the point and other end open to atmosphere. Simple manometers are further sub-divided into piezometers, U-tube manometers. Differential manometers measure pressure difference between two points either in the same pipe or different pipes at same or different level. Both ends of differential manometers are connected to the points, of which pressure difference is required.

2.6.1.1 Simple Manometer:

(a) Piezometer:

It consists of a glass tube, one end of which is open to atmosphere while the other end is connected to the point, the pressure at which is to be measured. The piezometer can be used to measure positive pressure or negative pressure as shown in figure 2.7. Due to pressure at the point, the liquid level rises or falls in the piezometer till the equilibrium is reached. If h is the rise in liquid level in the piezometer above the centre of pipe carrying liquid, M then pressure P at point M.

$$P = \gamma . H$$

where γ is specific weight of the liquid flowing through pipe.

Similarly, the negative pressure

$$p = - \gamma . h$$

The piezometer cannot be used for high pressures as well as gas pressure cannot be measured.

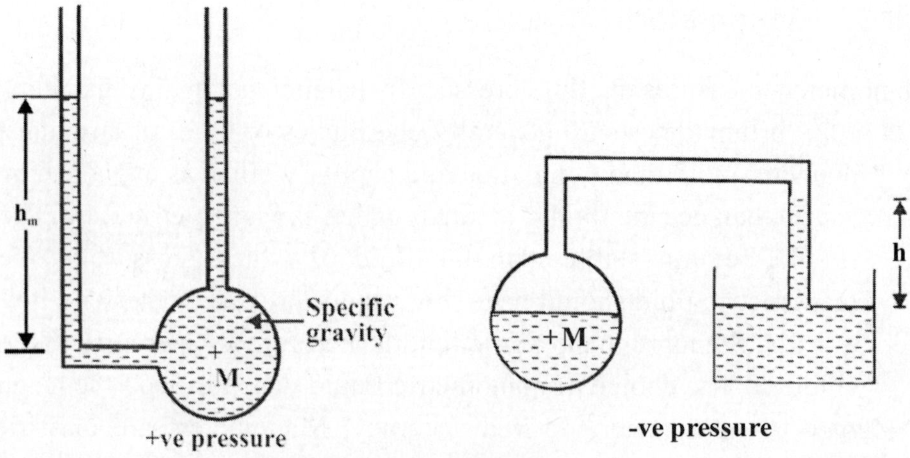

Fig. 2.7 Piezometer

(b) U-tube Manometer:

A tube bent in the shape of English letter 'U' is used to measure pressure at a point. One end of manometer is open to the atmosphere while the other end is connected to the point at which pressure is measured (Generally, centre of circular pipe). The rise or fall of manometric liquid with respect to point M gives the positive or negative pressure. For small pressures, same liquid which is flowing in the pipe can be used as manometric liquid. However, for large pressures heavier liquid like mercury is used as 'Manometric Liquid'. Fig. 2.8(a) shows arrangement of simple U tube manometer for measurement of positive pressure using the same liquid as in pipe.

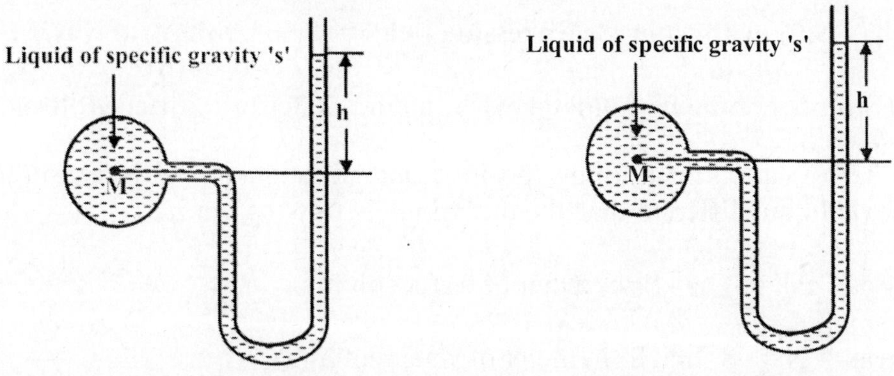

Small +ve pressure
Pressure at M = s.h. m of water
(a)

Small -ve pressure
Pressure at M = -s.h. m of water
(b)

Fig. 2.8 Simple 'U' tube manometers with same liquid in pipe and manometer

Pressure at M = h meter of manometric liquid of specific weight γ.

= S . h meter of water column where S is specific gravity of manometric liquid.

Pressure at M = S . h meter of water column. (2.11)

Fig. 2.8 (b) shows negative pressure measurement using U tube manometer using the same liquid as in pipe.

Pressure at M = -S . h meter of water column. (2.12)

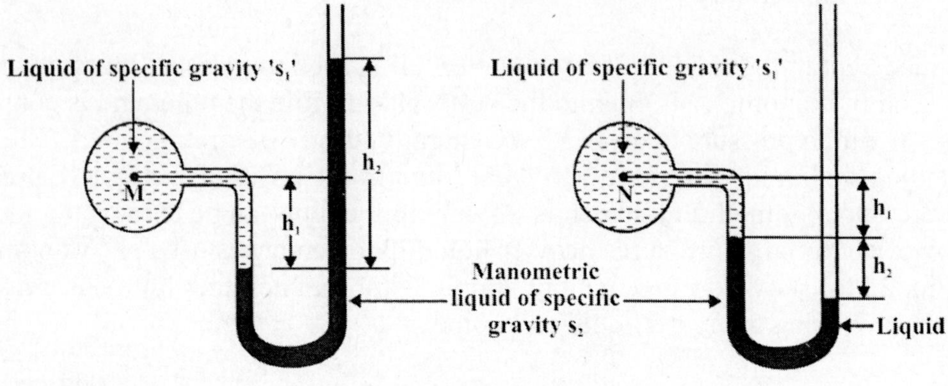

$h_M = S_2h_2 - S_1h_1$ m of water (+ve pressure) **(a)**   **$h_N = -S_2h_2 - S_1h_1$ m of water (-ve pressure)** **(b)**

Fig. 2.9 Simple U tube manometer with manometric fluid different than that of pipe

Figures 2.9 (a) and 2.9 (b) show positive and negative pressure measurement using a manometric liquid different than the one which is flowing in pipe.

Pressure at $M = S_2\,h_2 - S_1\,h_1$ meter of water column (2.13)

Pressure at $N = -S_2\,h_2 - S_1\,h_1$ meter of water column (2.14)

where S_1 is specific gravity of liquid flowing in the pipe and S_2 is specific gravity of manometric liquid.

2.6.1.2 Differential Manometer:

Figure 2.10 shows a differential manometer connected to measure pressure difference between 2 points M and N in pipes at different levels.

Let h_m and h_n be the pressure heads at 2 points m and n respectively in terms of water column. Then 1,

$$h_m + S_1\,h_1 + S_1\,h_2 + S_1\,h - S_2\,h - S_1\,h_2 = h_n$$

$$\therefore \quad h_m - h_n = h(S_2 - S_1) - S_1\,h_1 \text{ meter of water column.} \quad (2.15)$$

If the points M and N are at same level, then $h_1 = 0$

It may be noted that the above derived formulae for differential manometers are based on a simple procedure outlined below. There is no need to remember any formula. It can be derived as and when required very easily.

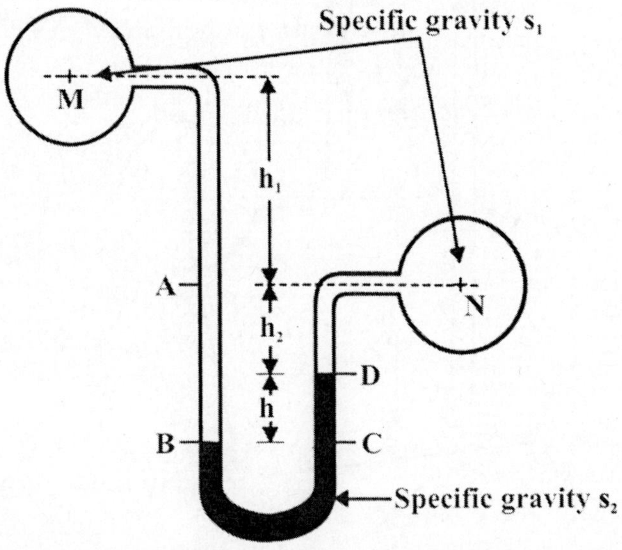

Fig. 2.10 Differential manometer

Procedure:

1. Draw a neat sketch of the system showing different liquid levels with their specific gravities.

2. In case of simple U tube manometer, start from open end (atmospheric pressure end) and in case of differential manometer, start from any end. Then go from level to level till the other end is reached. For decrease in the elevation, add the pressure head. For increase in the elevation, subtract the pressure head.

3. Express all heads in terms of equivalent water head.
For small pressure difference, an inverted U tube manometer with a lighter manometric liquid like air, carbon tetrachloride is used. A large deflection in the manometer is observed due to lighter manometric liquids. Figure 2.11 shows an inverted U tube manometer.

$$h_m - S_1 h_1 - S_1 h_2 - S_1 h + S_2 h + S_1 h_2 = h_n$$

$$\therefore \qquad h_m - h_n = h(S_1 - S_2) + S_1 h_1 \text{ meter of water column} \qquad (2.16)$$

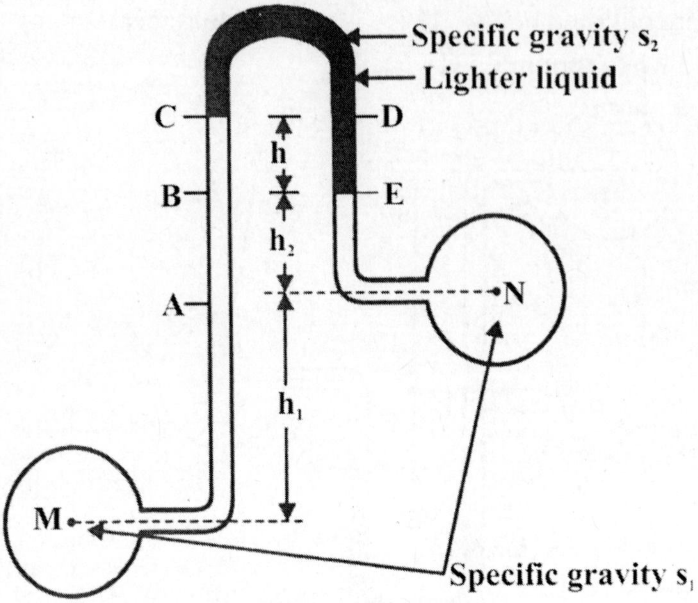

Fig. 2.11 inverted U-tube manometer

2.6.1.3 Sensitive Manometers:

While measuring deflections in simple or differential manometers many times readings cannot be measured accurately as the meniscus lies in between two millimetre marking and least count of the scale is 1 mm. In order to measure such deflections correctly sensitive manometers are used.

(a) Inclined tube manometer:

The deflection in the limb can be enlarged by making the mercury limb inclined at certain angle 'θ' with horizontal and pressure head can be measured accurately.

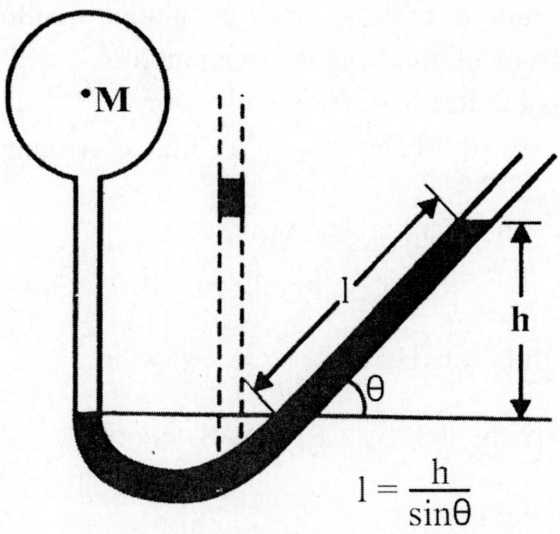

Fig. 2.12 shows an inclined tube manometer

$$l = \frac{h}{\sin\theta}, \text{ as } \sin\theta < l,\ l > h \tag{2.17}$$

(b) Single tube manometer or well type manometer:

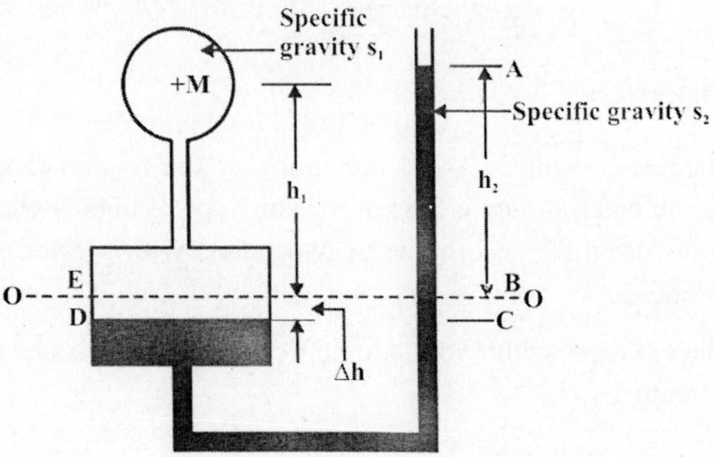

Fig. 2.13 Single tube manometer

In this type of manometer, a large reservoir, whose area is about 100 times of the area of the tube is introduced in one of the limbs of the manometer. Due to large area, the deflection in this limb is negligible as compared to the other limb of smaller area, through the volume of liquid displaced is same (direct application of Pascal's law)

As a result, only one reading of pressure head in smaller area limb is considered enough for pressure measurement, thus making manometer a quick instrument. However, by considering the small deflection in large reservoir, pressure measurement can be done more accurately. Figure 2.13 shows such a single tube manometer.

By neglecting the deflection in large reservoir

$$0 + S_2\, h_2 - h_1\, S_1 = h_m \quad (2.18)$$

If a small deflection of Δh is considered in large reservoir

$$0 + S_2\, h_2 + S_2\, \Delta h - S_1\, \Delta h - S_1\, h_1 = h_m$$

$$\therefore \quad h_m = S_2\, h_2 + (S_2 - S_1)\, \Delta h - S_1\, h_1 \quad (2.19)$$

However $\quad A\, .\, \Delta h = a.h_2 \quad$ (volume displaced by fluids is same)

$$\therefore \quad hm = S_2\, h_2 + (S_2 - S_1)\, x\, \frac{a}{A}\, h_2 - S_1\, h_1$$

$$\therefore \quad h_m = [S_2 + (S_2 - S_1)\, \frac{a}{A}]\, h_2 - S_1\, h_1 \text{ meter of water column} \quad (2.20)$$

(c) Micromanometer:

By introducing large reservoirs in both the limbs of the manometer, the reading are magnified and the accuracy is increased. This type of manometer which is used to measure pressure difference between two points with greater accuracy is called micromanometer.

Consider a micromanometer with two manometric fluids of specific gravity S_2 and S_3 as shown in figure 2.14.

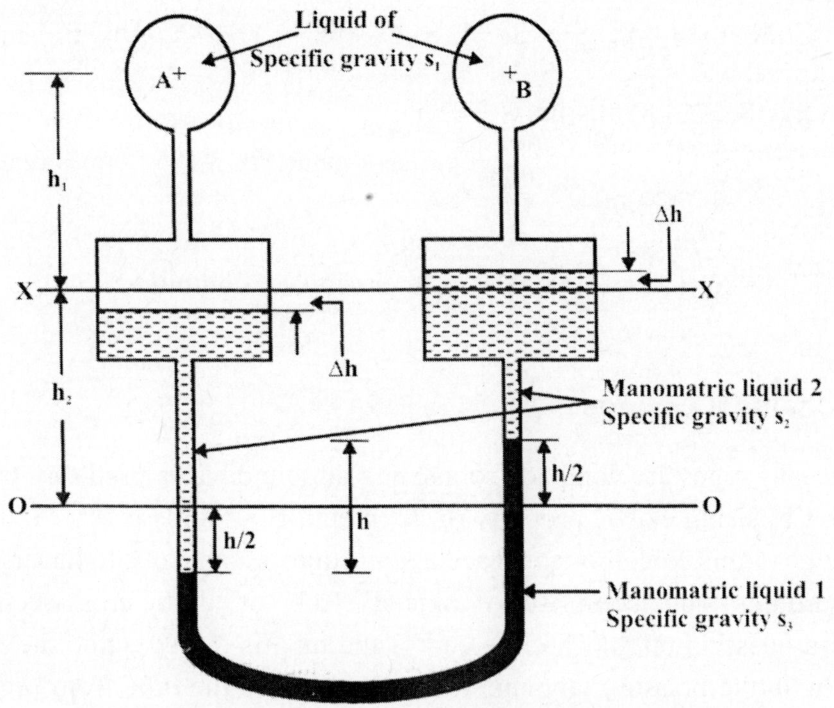

Fig. 2.14 Micromanometer

Initial level of heavier manometric liquid (S_3) is OO, while that of lighter liquid (S_2) is XX. The portion above XX is occupied by liquid (S_1) whose pressure is to be measured.

Let 'A' be the area of reservoir and 'a' be the area of tube.

$$\frac{A}{a} = 100$$

Volume displaced in tank = Volume displaced in tube

$$\therefore \quad A\,\Delta h = a.\frac{h}{2}$$

$$\therefore \quad \Delta h = \frac{a}{A}\frac{h}{2}$$

$$\therefore \quad 2\,\Delta h = \frac{a}{A}\,h$$

(Fall of Δh in left limb reservoir and rise of Δh in right limb reservoir will cause fall of $\frac{h}{2}$ in left tube and rise of $\frac{h}{2}$ in right tube)

$$h_A + S_1 h_1 + S_1 \Delta h + (h_2 - \Delta h) S_2 + S_2 \frac{h}{2} - S_3 h - (h_2 - \frac{h}{2}) S_2 - S_2 \Delta h - S_1 (h_1 - \Delta h) = h_B$$

$$h_A + S_1 (2 \Delta h) - S_2 (2 \Delta h) + S_2 h - S_3 h + h_B$$

$$h_A - h_B = S_3 h - S_2 h + (S_2 - S_1) \times \frac{a}{A} h$$

$$h_A - h_B = h \left\{ S_3 - S_2 \left(1 - \frac{a}{A}\right) - S_1 \frac{a}{A} \right\} \text{ meter of water column} \quad (2.21)$$

2.6.2 Mechanical Gauges:

Mechanical gauges are compact, robust and simple devices used to measure pressure at a point by using elastic property of the metal. Bourdon's pressure gauge is widely used gauge of this type, though its accuracy is questionable due to its larger least count. The Bourdon's gauge comprises of a metallic tube of elliptic cross-section bent in the form of a question mark. The free end of the tube is closed while the fixed end is in contact with the pressure tapping. When fluid enters the tube from pressure tapping, the closed end moves due to change in the cross-section of the tube from elliptical to circular. The movement of the closed end is transferred to the rack and pinion arrangement 'R' through a link 'L' which ultimately moves the pointer on the calibrated dial D (Refer figure 2.15). This arrangement can measure positive or negative pressure.

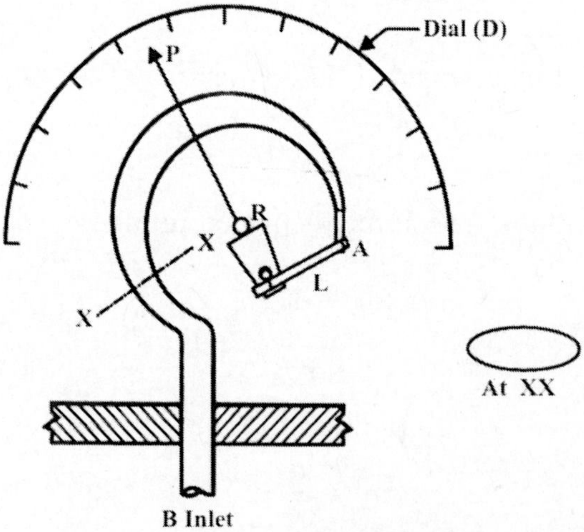

Fig. 2.15 Bourdon gauge

The pointer will move clockwise for positive pressure and anticlockwise for negative pressure. Different material of tube can be used for different pressure range. The gauge used to measure negative pressure is known as vacuum gauge. The calibration is generally made in kg/cm^2 or 'mm' of Hg.

2.6.2.1 Difference Between Mechanical Gauges and Manometers

Sr. No.	Mechanical gauge	Manometer
1.	Strong, portable, easy to handle.	Bulky, delicate and cannot be handled easily.
2.	Suitable for high pressure.	Can measure small pressures.
3.	Accurate at the centre of range.	Mostly accurate throughout the range.
4.	Sensitivity cannot be adjusted.	Sensitivity can be increased.
5.	Best for quick readings which may not be very accurate.	Best for accurate readings which may take some time to read.

2.7 Hydrostatic Forces on Surfaces

When a fluid comes in contact with any surface either plane or curved, it exerts a force on it. Obviously, this force is due to the pressure acting on the surface, which varies with depth of flow as explained in earlier sections. The total force acting on surface in contact of fluid is therefore termed as total pressure. When the fluid is at rest tangential forces (shear) are absent. As a result of which the total pressure acts normal to the surface on which it acts. Any force is described by three parameters, magnitude, direction and point of application. The magnitude of hydrostatic force or total pressure is governed by law of hydrostatics and direction is normal as explained previously. The point of application of the total pressure on the surface in contact is termed as centre of pressure. As explained earlier the pressure force varies with depth, the resultant of all such forces at different levels is the total pressure. The point of application i.e centre of pressure is found out using Varignon's theorem of moments which states that 'sum of moments of all the forces about a point is same as moment of the resultant about the same point'. In case of centre of pressure 'resultant moment' is total pressure and 'all the forces' are forces due to pressure at different levels. Thus, the concept of total pressure passing through centre of pressure is analogous with concept of weight of the body passing through centre of gravity.

2.8 Total Pressure of a Horizontal Plane Surface

Figure 2.16 shows a horizontal plane surface of area A is submerged in water at a depth 'h' below free water surface. The pressure intensity at all the points on this surface is constant as all points are at same depth below free water surface.

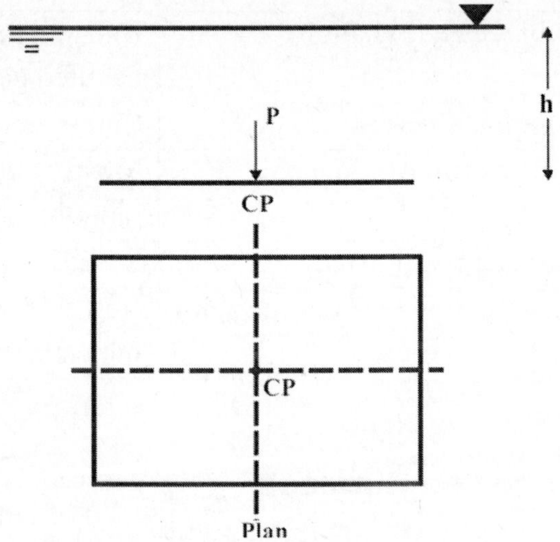

Fig. 2.16 Horizontal plane surface submerged in fluid

∴ $p = \gamma h$, where p is intensity of pressure and γ is specific weight of water.

Total pressure = Pressure intensity x area

∴ $P = (\gamma h) \times A$ (2.22)

It is also evident that this force will pass through centroid of area in the vertically downward direction.

2.9 Total Pressure on a vertical Plane Surface

Figure 2.17 shows a vertical plane surface of area A completely submerged in water.

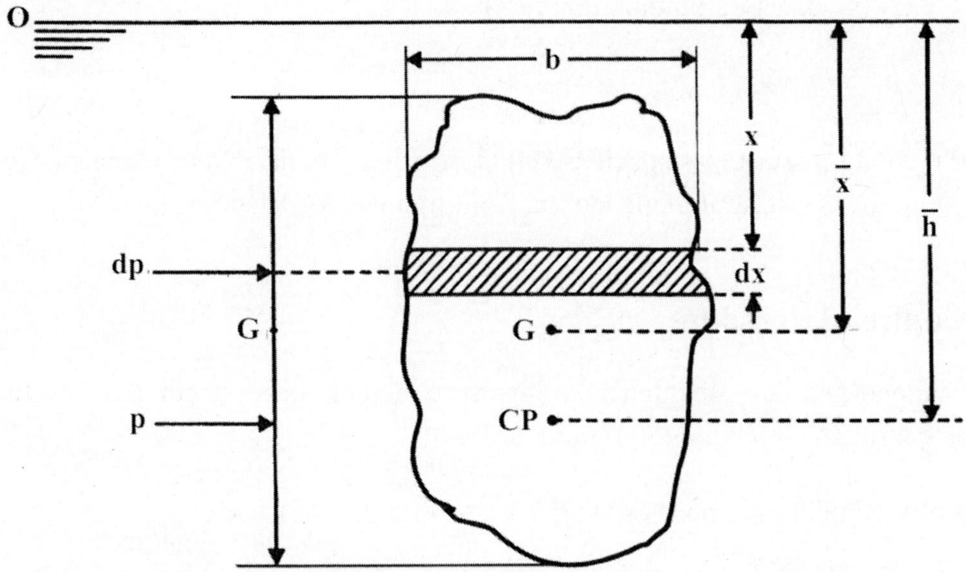

Fig. 2.17 Vertical plane surface submerged in fluid

- The centroid of plane surface is at a distance of $\bar{X}$ from free water surface OO.
- Pressure intensity varies over the plane surface (along the depth below the free water surface), therefore a rectangular strip of very small thickness 'dx' is considered at a distance 'x' from free water surface on which pressure intensity can be assumed to be constant.
- The width of strip is 'b'.

∴ Pressure intensity over strip of thickness dx = γx

∴ Total pressure on strip = Pressure intensity x area of strip

$$dp = (\gamma x)\,(bdx)$$

∴ Total pressure on entire plane surface $P = \int dp = \gamma \int x \,.\, (bdx)$

$\int x \,.\, (bdx)$ = Sum of moments of areas of all elemental strips about free water surface OO.

= Moment of the resultant area (total area) about free watersurface.

$$= A\bar{x} \text{ (Varignon's theorem)}$$

$$\therefore \quad P = \gamma A\bar{x} \tag{2.23}$$

Where P is total pressure, γ is specific weight of water, 'A' is area of plane surface and $\bar{x}$ is distance of the centroid of plane lamina from free water surface.

2.9.1 Centre of Pressure

For vertical plane surface the intensity of pressure increases with depth. Let $\bar{h}$ be the centre of pressure through which the total pressure P acts.

Moment of total pressure about OO = P $\bar{h}$ (refer figure 2.17)

Moment of total pressure acting on strip of thickness dx about OO = $(\gamma x)(bdx)\,x$

$\therefore$ Sum of moments of forces acting on all such strips about OO = $\gamma \int x^2 (bdx)$

Using Varignon's theorem of moment

$$\gamma \int x^2 (bdx) = P\bar{h}$$

$$\int x^2 (bdx) = \text{sum of second moment of areas about OO}$$

$$= \text{moment of inertia } I_o \text{ of plane surface about OO}$$

$$\therefore \quad P\bar{h} = \gamma I_o$$

$$\therefore \quad \bar{h} = \frac{\gamma Io}{P}$$

$$\therefore \quad \bar{h} = \frac{\gamma Io}{\gamma A\bar{x}^2} = \frac{Io}{A\bar{x}^2} \tag{2.24}$$

Using parallel axis theorem

$$I_o = I_G + A\bar{x}^2$$

i.e. moment of inertia about an is equal to sum of moment of inertia about a centroidal parallel axis in the same plane (I_G) and product of area and square of the distance between the two parallel axes ($A\bar{x}^2$)

$$\therefore \qquad \bar{h} = \frac{I_G + A\bar{x}^2}{A\bar{x}}$$

$$\therefore \qquad \bar{h} = \bar{x} + \frac{I_G}{A\bar{x}} \qquad (2.25)$$

As $\frac{I_G}{A\bar{x}} > 0$, $\bar{h} > \bar{x}$ i.e. centre of pressure ($\bar{h}$) will always lie below centre of gravity ($\bar{x}$).

For greater depths, centre of pressure approaches centre of gravity as $\bar{x}$ is large and $\frac{I_G}{\bar{x}}$ becomes small.

2.10 Total Pressure on Inclined Plane Surface

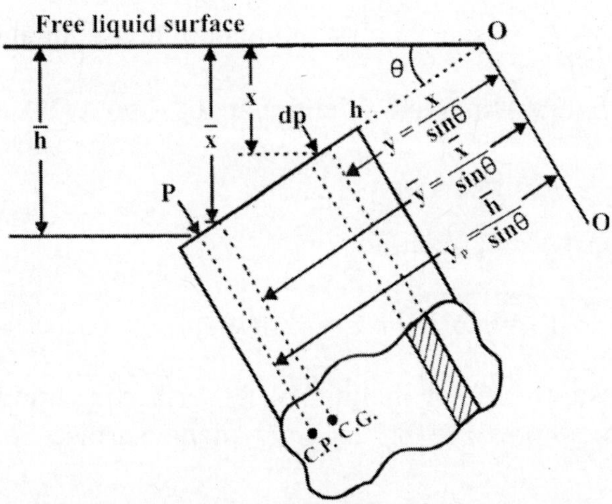

Fig. 2.18 Total pressure on inclined plane surface

Figure 2.18 shows an inclined plane surface of area A completely submerged in water. The angle of inclination of the place surface with the free liquid surface is θ. The plane of lamina when extended meets an axis 'OO' passing through O and perpendicular to the plane of paper.

Let $\bar{x}$ be the distance of centroid of plane surface vertically below free liquid surface. $\bar{y}$ be the inclined distance of centroid from OO along the inclined plane = $\frac{\bar{x}}{\sin\theta}$

Let $\bar{h}$ be the depth of centre of pressure vertically below free liquid surface and y_p be the inclined distance of centre of pressure from OO = $\frac{\bar{h}}{\sin\theta}$

Consider a strip of area dA at a distance x vertically from free water surface.

The intensity of pressure on strip = γx.

$\therefore$ Total pressure on strip $dp = \gamma x\,(dA)$

$\therefore$ $dp = \gamma\, y \sin\theta .\, dA$

$\therefore$ Total pressure on entire plane surface = $\gamma \sin\theta \int y\, dA$

$\int y\, dA$ = sum of First moment of areas about OO

= moment of entire area A about OO

= $A\bar{y}$

$\therefore$ $P = \gamma \sin\theta\, A\bar{y}$

$\therefore$ $P = \gamma A\bar{x}$ ($\because \bar{x} = \bar{y}\sin\theta$)

Thus, the total pressure acting on inclined plane surface is same as total pressure on vertical surface or total pressure on inclined plane surface is independent of angle of inclination θ.

2.10.1 Centre of Pressure

Moment of total pressure acting on strip about OO = $\gamma\, y \sin\theta\, dA \,.\, y$

$\therefore$ Moment of all such strips about OO = $\gamma \sin\theta \int y^2\, dA$.

$\int y^2\, dA$ = sum of second moment of areas about OO.

= Moment of inertia of the plane surface about OO i.e. I_o.

Using Varignon's theorem,

$$P \cdot y_p = \gamma \sin\theta \, I_o$$

$$I_o = I_G + A\,\bar{y}^2$$

$$y_p = \frac{\gamma \sin\theta\,(I_G + A\bar{y}^2)}{\gamma A \bar{y} \sin\theta} = \bar{y} + \frac{I_G}{A\,\bar{y}}$$

$$\bar{h} = \bar{x} + \frac{I_G \sin^2\theta}{A\,\bar{x}} \quad \left(\because y_p = \frac{\bar{h}}{\sin\theta} \text{ and } \bar{y} = \frac{\bar{x}}{\sin\theta}\right) \qquad (2.26)$$

2.11 Total Pressure on Curved Surface

Figure 2.19 shows a completely submerged curved surface under water. In case of curved surface, the total pressure is a vector sum of horizontal and vertical components P_H and P_V respectively as the elementary forces do not form a parallel force system as in plane surfaces. Consider an elementary area dA situated at a depth h below free water surface. the elementary area is magnified in Fig. 3.11.2.

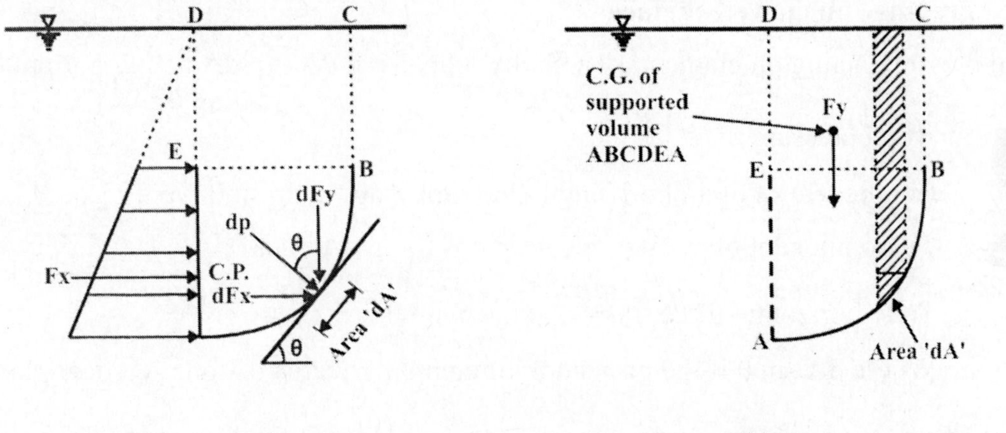

(a) **(b)**

Fig. 2.19 Total pressure on curved surface

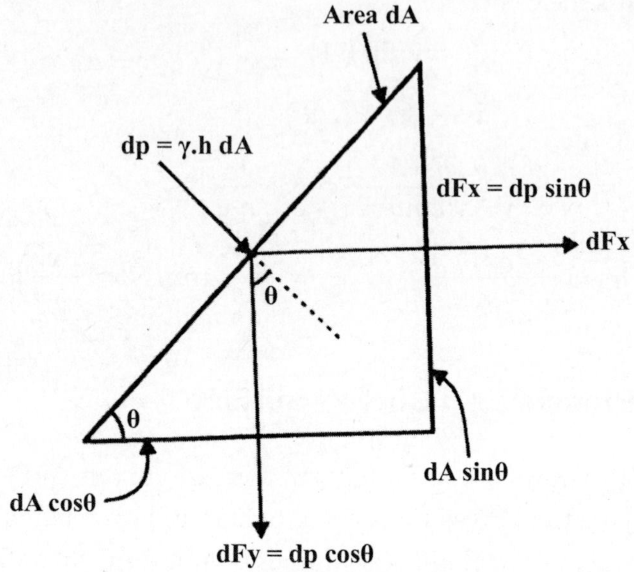

Fig. 2.20 Elementary area dA

Total pressure on curved surface

Total pressure acting on elementary area dA = dp = γ h dA = p dA acting normal to dA.

$\therefore$ Total pressure $P = \int \gamma\, h\, dA$

Component of dp in horizontal direction = dp_H = dp sinθ = p . dA sinθ.

And vertical component of dp = dp_V = dp cos θ = p dA . cos θ

Total horizontal force $P_H = \gamma \int h\, dA \sin\theta$

Where dA sin θ is the projection of elementary area dA on a vertical plane.

$\therefore$ $P_H = \int \gamma\, h\, dA \sin\theta$ = total pressure on projected area of the curved surface on vertical plane PH will act at centre of pressure. (2.27)

dA cos θ = horizontal projection of elementary area dA.

$\therefore$ $P_V = \int \gamma\, h\, dA \cos\theta$ = Weight of the liquid lying above curved surface in Area ABCDEFA

Thus, the vertical component of total pressure is the weight of liquid lying in the portion extending above curved surface up to free surface and acting through centroid of area ABCDEFA. P_V will act in vertically downward direction if supported liquid is real otherwise it will act vertically upwards.

The total pressure $P = \sqrt{P_H^2 + P_V^2}$

$$= \tan^{-1} \frac{P_V}{P_H} \quad (2.28)$$

Point of application of total pressure P on curved surface may be determined by extending the line of action of force P to meet the surface.

2.12 Solved Examples

Ex.2.1:	Mercury barometer reads 720 mm at top of a mountain. Calculate height of mountain if atmospheric pressure at the bottom of mountain is 109 kPa. Assume density of air 1.26 kg/m3 constant If the plate remains equidistant from the two surfaces. **Solution:** Pressure at top $P = \gamma$ h = 13.6 x 9.81 x 0.72 $= 96.06 \frac{kN}{m2}$ or kPa Pressure at bottom = 109 kPa ∴ Difference = 12.94 kPa $P = \gamma h$ (for air) 12.94 x 103 = 1.26 x 9.81 x h h = 1046.87 m

	∴ height of mountain = 1046.87 m

Ex. 2.2 :	Mass density (kg/m^3) of a liquid varies as p = (1000 + 0.008 h) where 'h' is depth (m) below free surface of liquid. Determine depth at which gauge pressure would be 100 kPa. **Solution :** $p = 1000 + 0.008\ h$ ∴ $\gamma = 9810 + 0.07848\ h$ $\frac{dp}{dh} = \gamma$ ∴ $\int_{h_1}^{h_2} P_1 - P_2 = \int_{h_1}^{h_2} \gamma dh = \int (9810 + 0.07848\ h)\ dh$ $P_2 = 0$ since gauge pressure is asked, pressure on Surface = 0 (atmp) ∴ $100 \times 10^3 = 9810\ h + 0.03924\ h^2$ where $h = h_2 - h_1$ ∴ $h^2 + 250000\ h - 2548420 = 0$ $H = 10.193$ m

Ex. 2.3 :	What depth of oil, sp. Gravity 0.8 will produce a pressure of 120 kN/m^2 ? What would be corresponding depth of water? **Solution** : $P = \gamma\ h$ ∴ $120 \times 10^3 = 0.8 \times 9810 \times h$ ∴ $h_{oil} = 15.29$ 'm' of oil

	$H\omega =$ S_{oil} x h_{oil} = 0.8 x 15.29 = 12.23 'n' of water (Alternatively $h\omega = \frac{P}{h\omega} = \frac{120 \, x \, 10}{9810}$ = 12.23 'm' of water)

Ex. 2.4 :	A bourdon tube is connected to a tank in which the pressure is 276 kPa above atmospheric at the gauge connection. If the pressure in the tank remains unchanged but the gauge is placed in a chamber where the air pressure is reduced to a vacuum of 635 mm of mercury. What gauge reading will be expected? (The gauge connection is not shifted) **Solution:** $\frac{276}{9.81}$ = 28.134 m of water Pressure reduced to 635 mm of Hg = 0.635 x 13.6 = 8.363 of water Total pressure = 28.134 = 19.498 m of water Gauge reading = 19.498 x 9.81 = 191.275 kPa

Ex. 2.5 :	Two pressure tanks are built one inside the other. A bourdon gauge M connected to the inner tank reads 20 kPa. Another bourdon gauge 'N' connected to the outer tank reads 35 kPa. An aneroid barometer reads 750 mm of Hg. Calculate the absolute pressure recorded at M and N in mm of mercury. **Solution:** Pressure recorded by M = 35,000 + 20,000 = 55,000 N/m^2

	Pressure recorded at M = $\frac{55,000}{9810 \times 13.6}$ = 0.412 m mercury Absolute pressure recorded by M = $\frac{750}{1000}$ + 0.412 = 1.162 m of mercury = 116.22 cm of mercury Pressure recorded by N = 35,000 N/m^2 Pressure recorded by N = $\frac{35,000}{9810 \times 13.6}$ = 0.262 m mercury Absolute pressure recorded at N = $\frac{750}{1000}$ + 0.262 = 1.012 m of Hg = 101.2 cm of Hg.

Ex. 2.6 :	A U-tube manometer Fig. Ex. 2.6 measures the pressure difference between points A and B on a liquid of density ρ_1. The U-tube contains mercury of density ρ_2. Calculate the difference of pressure if a = 1.5, b = 0.75 m and h = 0.5 m if the liquid in A and B is water and $\rho_2 = 13.6\ \rho_1$.

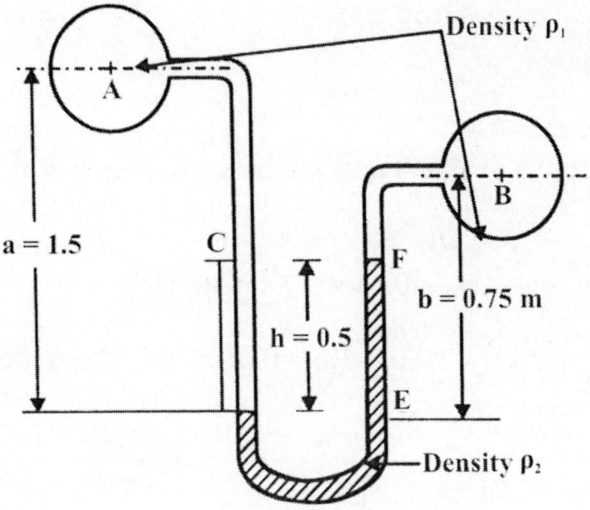

Fig. Ex. 2.6

Solution :

$$H_A + 1.5 - 13.6 \times 0.5 - 0.25 = H_B$$

$H_A - H_B = 5.55$ 'm' of water column.

Ex. 2.7 : Determine $P_A - P_B$ as shown in Fig. Ex. 2.7

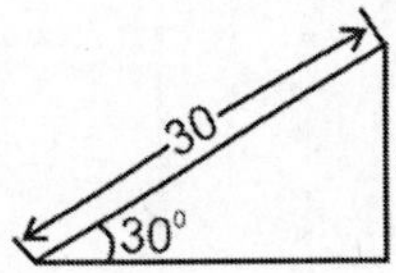

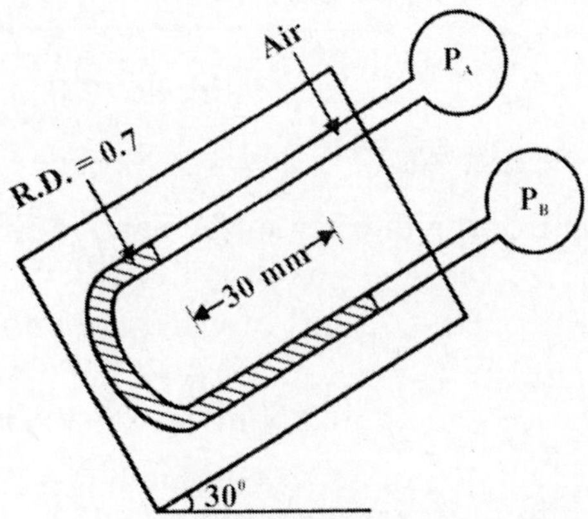

Fig. Ex. 2.7

Solution :

	$\sin 30 = \dfrac{h}{0.03}$ $\therefore \quad h = 0.03 \sin 30$ Pressure due to air column will be negligible, as Relative density of air is very small. $H_A - 0.03 \times 0.7 \sin 30^0 = H_B$ $\therefore \quad H_A - H_B = 0.0105$ 'm' of water column. $\therefore \quad P_A - P_B = 0.0105 \times 9810 = 103.005$ N/m^2

Ex. 2.8 :	Calculate intensity of pressure at points A, B, C and D as shown in Fig. Ex. 2.8. 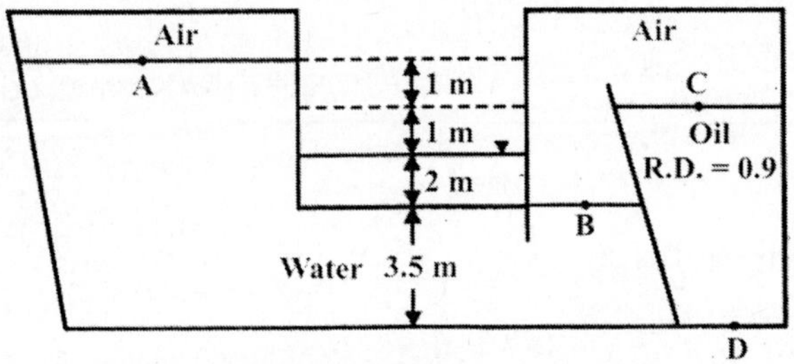 **Fig. Ex. 2.8** **Solution :** 1. Starting from free water surface and moving towards point A. $H_A = -2$ 'm' of water column. $\therefore \quad P_A = -19.62$ kN/m^2 $(P_A = \gamma_W H_A)$ 2. Starting from free water surface and moving towards B. $H_B = 2$ 'm' of water column $= 19.62$ kN/m^2

	3. Pressure at B = Pressure at C = 19.62 (weight of air is negligible) 4. $H_C + 6.5 \times 0.9 = 2 + 5.85 = 7.85$ 'm' of water column $= 77\ kN/m^2$

Ex. 2.9 : The diameter of two limbs A and B of a U-tube manometer are 16 mm and 4 mm respectively. A is filled with a liquid of specific gravity of 1.25 and B with another liquid of specific gravity 0.8. The two liquids do not mix. Determine the pressure to be applied to the surface of the heavier liquid in order to raise the level in the other limb by 10 mm.

Solution :

Pressure applied on left limb (S = 1.25) is same as pressure on right limb (S = 0.8)

$$\therefore \quad p = \gamma_1 h_1 = \gamma_2 h_2$$

Let the level of liquid of specific gravity 1.25 decreases by h_1 m i.e. pressure head

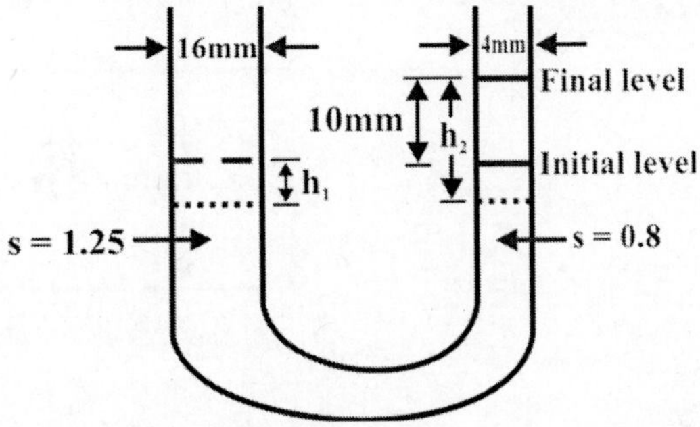

Fig. Ex. 2.9

of 1.25 m of liquid of specific gravity 1.25.

Due to this liquid level in the other limb rises by 10 mm, above original level and by h_2m above new level of liquid in the first limb (s = 1.25). As pressure applied on first limb is balanced by adjustment of liquid level in the second limb.

$$P = \gamma_1 h_1 = \gamma_2 h_2$$

but,

$$h_2 = h_1 + 0.01$$

$$1.25 \times 9.81 \times h_1 = 0.8 \times 9.81\,(h_1 + 0.01)$$

$$\therefore \quad h_1 = \frac{0.008}{0.45} = 0.0177 \text{ 'm' of liquid of specific gravity } 1.25$$

$$\therefore \quad p = \gamma_1 h_1 = 1.25 \times 9.81 \times 0.0177 = 0.218 \text{ kN/m}^2$$

Ex. 2.10 : A closed rectangular tank of cross sectional area 1m x 0.6 m has 1 m height. It contains water upto depth of 0.6 m and remaining space above contains air under such a pressure that total load on the bottom of tank is 9.532 kN. Determine pressure of air.

Solution :

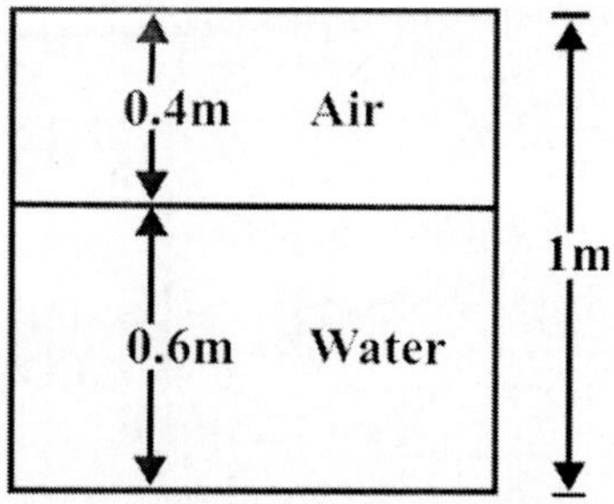

Fig. Ex. 2.10

$$P_{total} = P_a + P_w$$

$$P_{bottom} = \frac{9.532}{1 \times 0.6} = 15.89 \text{ kN/m}^2$$

$$P_w = 0.6 \times 9.81 = 5.886 \text{ kN/m}^2$$

$$P_{air} = 10.004 \text{ kN/m}^2$$

Ex. 2.11 :	A vessel of 5 cm^2 cross sectional area and 1.5 m height is filled with water upto a height of 1 m and remaining with oil of specific gravity 0.8. The vessel is open to atmosphere. Calculate the gauge and absolute pressure on the base of vessel in terms of water head, oil head and N/m^2 given the atmosphere as 1.013 bar. Also calculate net force exerted at the base of vessel. **Solution :** $P_{base} = P_{oil} + P_{water} = 0.5 \times 0.8 \times 9.81 + 1 \times 9.81$ $= 13.734\ kN/m^2$ (gauge) Atmosphere $= 1.013$ bar $= 1.013 \times 10^5\ N/m^2$ $= 101.3\ kN/m^2$ $\therefore$ Total force $= 101.3 + 13.734$ $= 115.034\ kN/m^2$ (absolute) $\therefore$ Total force $= 115034\ N/m^2$ (absolute) In terms of water head, (convert oil head to water head) $H_w = s_w h_w + s_{oil} h_{oil}$ $= 1 \times 1 + 0.5 \times 0.8$ $= 1.4$ 'm' of water column In terms of oil head (convert water head to oil head) $\therefore \gamma_w h_w = s_{oil} h_{oil}$ $\therefore h_{oil} \times s_{oil} = h_w \qquad \therefore h_{oil} = \frac{1}{0.8} = 1.25$

$$\therefore \quad \text{Total head,} \quad h_{oil} = \frac{h_w}{S_o} + h_{oil}$$

$$\therefore \quad h_{oil} = 1.25 + 0.5 = 1.75 \text{ 'm' of oil (gauge).}$$

Ex. 2.12 : A square plate of diagonal 1.5 m is immersed in water with its diagonal vertical and upper corner 0.5 m below the free surface of water. Calculate the depth of C.P. on the plate from free surface of water and hydrostatic force resulting on the plate in kN.

Solution :

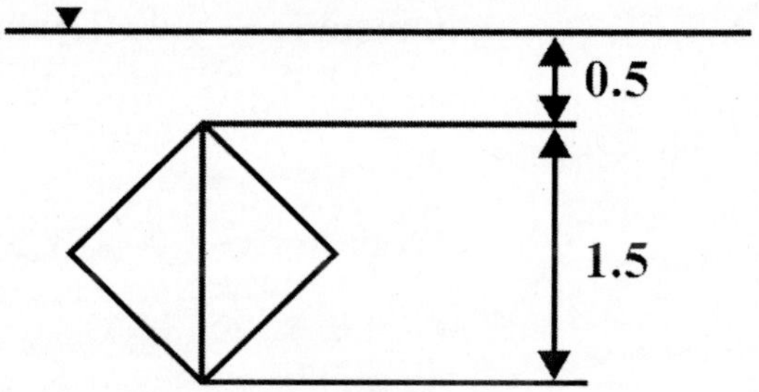

Fig. Ex. 2.12

$$\text{Side of square} = \frac{1.5}{\sqrt{2}} = 1.06 \text{ m}$$

$$\text{Area} = 1.125\text{m}^2$$

$$P = 9.81 \times 1.125 \times 1.25 = 13.800 \text{ kN}$$

$$\bar{h} = \bar{x} + \frac{I_G \sin^2\theta}{A\,\bar{x}}$$

$$= 1025 + \frac{\frac{1.5 \times 0.75^3}{12} \times 1}{1.125 \times 1.25} \qquad (\because \theta = 90^0)$$

$$\bar{h} = 1.325 \text{ m}$$

Ex. 2.13 :	A triangular plate of 1 m base of 1.5 m latitude is immersed in water. The plane of plate is inclined at 30^0 with free water surface and base is parallel to and at a depth of 2 m from water surface. Find T.P. and C.P.

Solution :

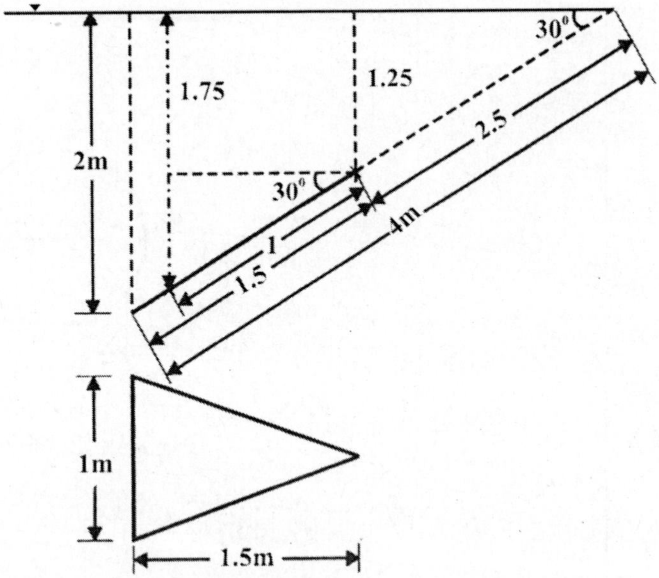

Fig. Ex. 2.13

$$T.P. = \gamma\, A\, \bar{x}$$

$$= 9.81 \times \frac{1}{2} \times 1 \times 1.5 \times 1.75$$

$$= 12.88 \text{ kN}$$

$$C.P. = \bar{x} + \frac{I_G \sin^2\theta}{A\,\bar{x}}$$

$$= 1.75 + \frac{\frac{1 \times 1.5^3}{36} \times 0.5^2}{\frac{1}{2} \times 1.5 \times 1.5}$$

$$= 1.77 \text{ m}$$

Ex. 2.14 : A circular plate 1.2 m diameter is placed vertically in water so that the centre of plate is 2 m below free surface. Determine the total pressure on the plate and depth of C.P.

Solution :

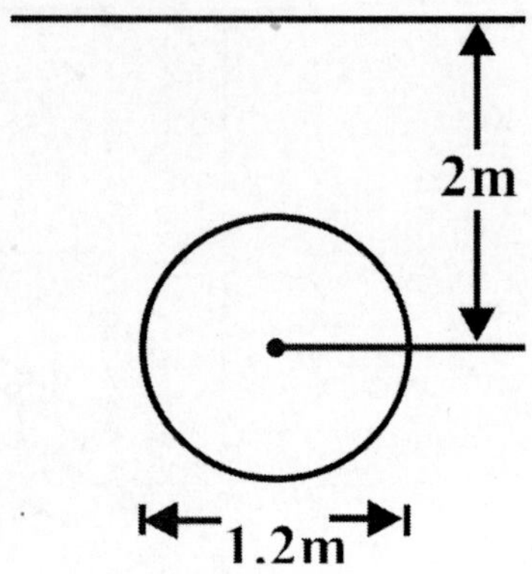

Fig. Ex. 2.14

$$p = 9.81 \times \frac{\pi}{4} \times 1.2^2 \times 2 = 22.19 \text{ kN}$$

$$\bar{h} = 2 + \frac{\frac{\pi}{64} \times 1.2^4}{\frac{\pi}{4} \times 1.2^2 \times 2} = 2.045 \text{ 'm'}$$

Ex. 2.15 : A plate 4 m long and 2 m wide has a circular hole of 1 m diameter at its centre. The plate is completely immersed in water making an angle of 450 with free surface. Determine the hydrostatic load on one face of plate and C.P.

Solution :

P_1 = Total pressure with hole = $\gamma A \bar{h}$ $\quad (\bar{h} = 2 \sin 45 = 1.41\text{m})$

= 9.81 x 8 x1.41

= 110.66

$$\text{Centre of pressure } h_1 = 1.41 + \frac{\frac{2 \times 4^3}{12} \times \frac{1}{2}}{8 \times 1.41}$$

$= 1.88$ m

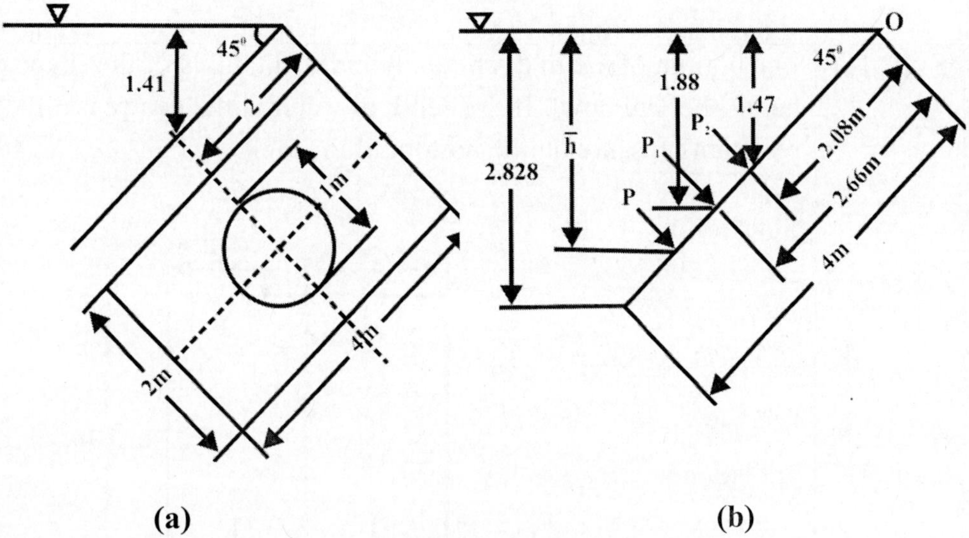

Fig. Ex. 2.15

$P_2 =$ Total pressure on hole

$= 9.81 \times \frac{\pi}{4} \times 1 \times 1.41 = 10.86$

$$\text{Centre of pressure } h_2 = 1.41 + \frac{\frac{\pi}{64} \times 1^4 \times \frac{1}{2}}{\frac{\pi}{4} \times 1 \times 1.41} = 1.47 \text{ m}$$

$\therefore \quad P = P_1 - P_2 = 110.66 - 10.86$

$= 99.8$ kN acting at $\bar{h}$ from free water surface

(Distance of centre of pressure along the plate are shown in Fig. Ex. 2.15(b))

Taking moments about 'O'

$$110.66 \times \frac{1.88}{0.707} - 10.86 \times \frac{1.47}{0.707} = 99.8 \times \frac{\bar{h}}{0.707}$$

Centre of pressure $\bar{h}$ = 1.92 'm'

Ex. 2.16: The bottom of a 1 m diameter cylindrical tank is of the shape of an inverted hemispherical bowl. If the depth of water at the centre is 3 m, find the resultant pressure on the bottom of the tank.

Solution :

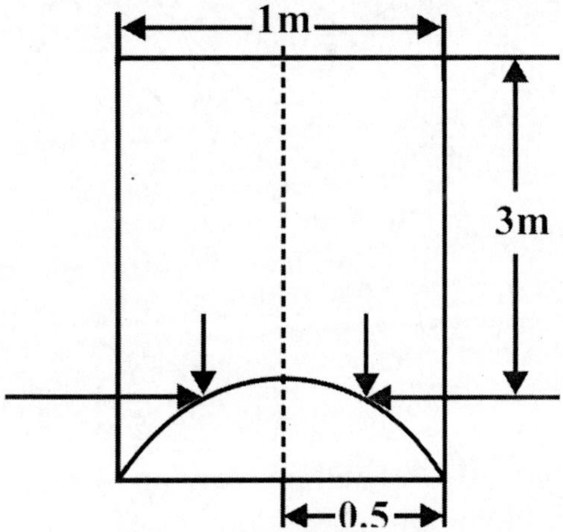

Fig. Ex. 2.16

Total pressure on half of cylinder will be resultant of vertical and horizontal component. But for entire cylinder, horizontal components will cancel and vertical components i.e. weight of water columns will add.

$\therefore$ Total pressure = weight of water acting on bottom of cylinder

$\therefore$ Total pressure = γ x volume

$$= 9.81 \times [\pi \times r^2 \times h - \frac{2}{3} \pi r^3]$$

	$= 9.81 \times [\pi \times 0.5^2 \times 3.5 - \frac{2}{3} \times \pi \times 0.5^3]$ $= 24.39\ kN/m^2$

Ex. 2.17:	A 1 m wide and 1.5 m deep rectangular plane surface lies in water in such a way that its plane makes an angle of 30^0 with the free water surface. Determine the total pressure and position of centre of pressure when the upper edge is 0.75 m below the free water surface. **Solution :** 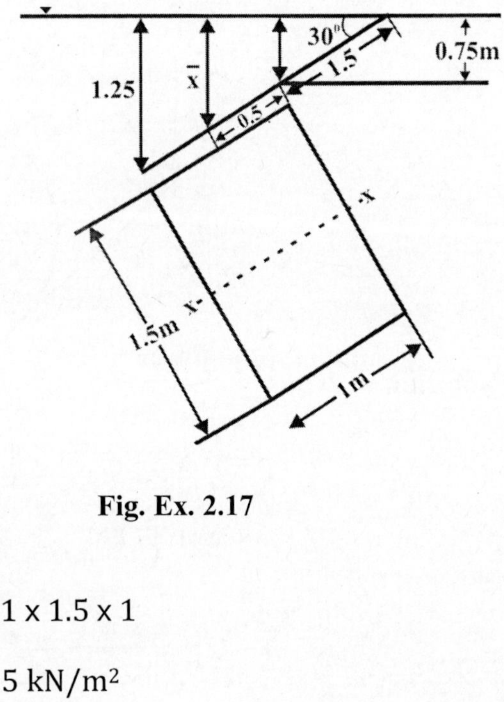 **Fig. Ex. 2.17** $p = \gamma A \bar{x}$ $= 9.81 \times 1 \times 1.5 \times 1$ $= 14.715\ kN/m^2$ $\bar{h} = \bar{x} + \frac{I_G \sin^2\theta}{A \bar{x}}$ $= 1 + \frac{\frac{1 \times 1.5^3}{36} \times 0.5^2}{1 \times 1.5 \times 1}$

	$\bar{h}$ = 1.046825 m

Ex. 2.18:	A circular plate of 4 m diameter is immerged in water such that its greatest and least depth below free surface of water is 5 m and 2 m respectively. Determine total pressure on one side of plate and position of centre of pressure.

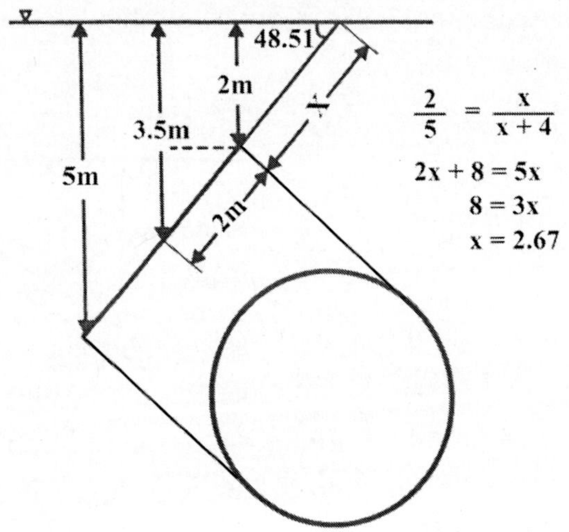

Solution :

$$p = \gamma\, A\, \bar{X} = 9.81 \times \frac{\pi}{4} \times 16 \times 3.5$$

$$= 431.47 \text{ kN}$$

$$\bar{h} = \bar{x} + \frac{I_G \sin^2\theta}{A\,\bar{x}} = 3.5 + \frac{\frac{\pi}{64} \times 4^4 \times 0.561}{\frac{\pi}{4} \times 16 \times 3.5}$$

$$= 3.66 \text{ 'm'}$$

Ex. 2.19:	A tank contains water upto a height of 0.5 m above the base. An immiscible liquid of specific gravity 0.85 is filled on the top of water upto 1 m height.

Calculate :

(i) Total pressure on one side of the tank.

(ii) The position of centre of pressure for one side of the tank, 2 m wide.

Solution :

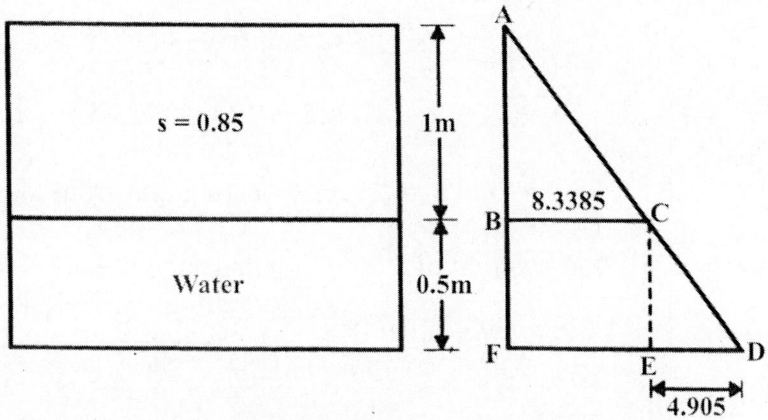

Fig. Ex. 2.19

As the liquids of different specific gravity are used, pressure, diagram will be used.

Intensity at 1 'm' = $p = \gamma_1 h_1 = 0.85 \times 9.81 \times 1$

$= 8.3385$ kN/m^2

Intensity at 0.5 m = $9.81 \times 0.5 = 4.905$ kN/m^2

P_1 = Pressure force exerted by prism abc

$= \frac{1}{2} \times 8.3385 \times 1 \times 2 = 8.3385$

P_1 acts through C.G. of pressure prism

$\therefore \quad \bar{h}_1 = \frac{2}{3} \times 1$ from 'A'

	$P_2 =$ Area (BCEF) x 2 = 8.3385 x 0.5 x 2 = 8.3385 at 1.25 'm' from A $P_3 = \frac{1}{2} \times 4.905 \times 0.5 \times 2$ $= 2.4525$ at $[1 + \frac{2}{3} \times 0.5] = 1.335$ from A ∴ Total force = 8.3385 + 8.3385 + 2.4525 = 19.1295 acting at $\bar{h}$ from 'A' To find position of P take moments about A $P\,\bar{h} = P_1\,\bar{h}_1 + P_2\,\bar{h}_2 + P_3\,\bar{h}_3$ $19.1295\,\bar{h} = 8.3385 \times \frac{2}{3} + 8.3385 \times 1.25 + 2.4525 \times 1.335$ $\bar{h} = 1$ 'm' below 'A'

Ex. 2.20:	A tank contains water upto a height of 1 m above the base. An immiscible liquid of specific gravity 0.8 is filled on the top of water upto a height of 0.5 m. Calculate the total pressure on one side of tank and locate position of centre of pressure for one side of the tank which is 2 m wide. Also plot pressure diagram. **Solution :**

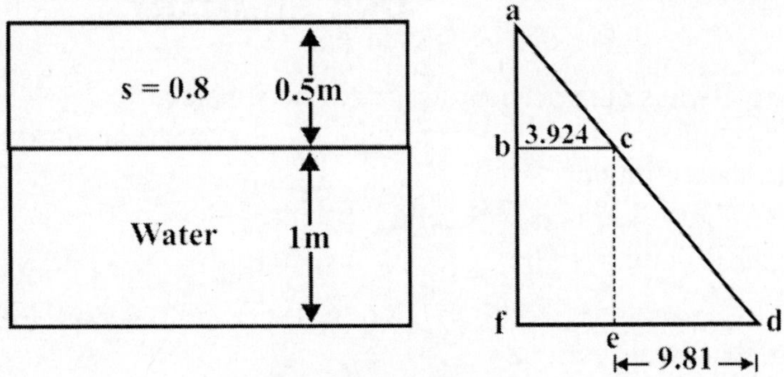

Fig. Ex. 2.20

'Intensity at 1 m $= P_1 = \gamma_1 h_1 = 9.81 \times 0.8 \times 0.5 = 3.924$ kN/m^2

Intensity at 1.5 m $= P_2 = 9.81 \times 1 = 9.81$

Pressure force P_1 $= \Delta\ abc = \frac{1}{2} \times 3.924 \times 0.5 \times 2$

$= 1.962$ acting at $\frac{2}{3} \times 0.5 = 0.355$ m from a

Pressure force P_2 $= \Delta\ ced = \frac{1}{2} \times 9.81 \times 1 \times 2$

$= 9.81$ actng at $0.5 + \frac{2}{3} \times 1 = 1.17$ m from a

Pressure force P3 $= \square\ bcef = 3.924 \times 1 \times 2$

$= 7.848$ acting at $0.5 + 0.5 = 1$ m from a

$\therefore$ Total force $= 1.962 + 9.81 + 7.848$

$= 19.62$ kN acting at $\bar{h}$ from a

Taking moments about 'a'

$$19.62 \times \bar{h} = 1.962 \times 0.335 + 9.81 \times 1.17 + 7.848 \times 1$$

$$\bar{h} = 1.0185 \text{ m}$$

UNIT SUMMARY

- Pressure always acts perpendicular to the surface.
- Pascal's law with derivation.
- Hydrostatics law $p = \gamma h$
- Piezometric head $= \frac{P}{\gamma} + z$
- Equivalent water column $h_{water} = S_{fluid} \times h_{fluid}$
- When pressure is measured above atmospheric pressure, it is called gauge pressure or positive pressure.
- When pressure is measured below atmospheric pressure it is called vacuum pressure or negative pressure.
- Absolute pressure is always positive as it is measured above absolute zero.
- Simple and differential 'U' tube manometer.
- Inverted 'u' tube manometer and micromanometer.
- Total pressure on a vertical plane surface (Derivation)

  $$p = \gamma A \bar{X}$$

  and centre of pressure $\bar{h} = \bar{x} + \frac{I_G}{A\,\bar{x}}$

- Total pressure on an inclined plane surface (Derivation)

  $$p = \gamma A \bar{x}$$

  and centre of pressure $\bar{h} = \bar{X} + \frac{I_G \sin^2\theta}{A\,\bar{x}}$

- Total pressure on an curved surface $p = \sqrt{P_H^2 + P_V^2}$ $\alpha = \tan^{-1}\frac{P_V}{P_H}$ (Derivation)

2.13 Exercise

2.13.1: Objective Questions:

1. The normal stress in a fluid will be constant in all directions at a point only if…

   (a) It is incompressible

   (b) It has uniform viscosity

   (c) It has zero viscosity

   (d) It is at rest.

   Ans: (d)

The atmospheric pressure with rise in altitude decreases…

   (a) Linearly

   (b) first slowly and then steeply

   (c) first steeply and then gradually

   (d) unpredictable.

   Ans: (b)

2. Mercury is often used in barometer because…

   (a) It is the best liquid

   (b) The height of barometer will be less

   (c) Its vapour pressure is so low that it may be neglected

   (d) Both (b) and (c).

   Ans: (d)

3. A pressure of 25 m of head of water is equal to…

   (a) 25 kN/m^2

   (b) 245.25 kN/m^2

   (c) 2500 kN/m^2

   (d) 2.5kN/m^2

Ans: (b)

4. Barometer is used to measure…

   (a) Pressure in pipes, channels etc

   (b) Atmospheric pressure

   (c) Very low pressure

   (d) Difference of pressure between two points.

   Ans: (b)

5. Which of the following manometer has highest sensitivity?

   (a) U-tube with water

   (b) inclined U-tube

   (c) U-tube with mercury

   (d) micro-manometer with water.

   Ans: (d)

6. Along the free surface in a liquid, pressure ________

   (a) Increases

   (b) Decreases

   (c) Remains constant

   (d) Not from above three options

   Ans: (c)

7. In a differential manometer a head of 0.6 m of fluid *A* in limb 1 is found to balance a head of 0.3 m of fluid *B* in limb 2. The ratio of specific gravities of *A* to *B* is

   (*a*) 2 (*b*) 0.5 (*c*) cannot be determined (*d*) 0.18

   Ans: (b)

8. The specific weight of a fluid is 20,000 N/m3. The pressure (above atmosphere) in a tank bottom containing the fluid to a height of 0.2 m is

(*a*) 40,000 N/m^2 (*b*) 2000 N/m^2 (*c*) 4000 N/m^2 (*d*) 20,000 N/m^2

Ans: (c)

10. Manometers are suitable for ________ pressure measurement.

(*a*) Low (*b*) High (*c*) Medium (*d*) Extreme high

Ans: (a)

11. If the density varies linearly with height the pressure will vary ________ with height.

(a)Linearly (*b*) in proportion (*c*) inversely (*d*) Exponentially

Ans: (d)

12. In micromanometer, the density difference between the filler fluid and the manometer fluid should be ________

(a)small (*b*) High (*c*) Medium (*d*) Extreme high

Ans: (a)

13. The pressure on the base of a liquid column will depend upon the shape of the column.

(a) Correct (*b*) Incorrect

Ans: (b)

14. For low pressure measurement a manometric fluid with low density will be better.

(a) Correct (*b*) Incorrect

Ans: (a)

15. The vacuum gauge reading will increase as the absolute pressure decreases.

(a) Correct (*b*) Incorrect

Ans: (a)

2.13.2: Theory Questions:

Q.1 What is Pascal's law? Derive it.

Q.2 What is meant by equivalent head?

Q.3 With the help of a neat sketch show that if a differential manometer is used to measure the pressure difference between two points, the deflection takes into account datum head of points also.

Q.4 Write short note micro manometer.

Q. 5 Explain with neat sketch the working of single column manometer.

Q. 6 Write a note of Bourdon's gauge.

Q. 7 What is meant by centre of pressure?

Q. 8 Derive an expression for the vertical distance between the centre of gravity and centre of pressure of a plane immersed surface.

$$\frac{\bar{h}=\bar{x}\ I_G}{A\ \bar{x}}$$

Q. 9 Derive the expressions for forces and their point of applications on curved surfaces.

Q. 10 What is a pressure diagram? What are its uses and limitations?

Q. 11 Derive expression for total pressure and centre of pressure on an inclined plane surface.

Q. 16 Derive an expression for the total pressure acting on plane surface kept in liquid at angle 'θ' with the free liquid surface. Also determine the location of centre of pressure.

Q. 20 Define total pressure and centre of pressure.

Q. 21 Derive the expression for determining the centre of pressure for an inclined triangular plane immersed in water.

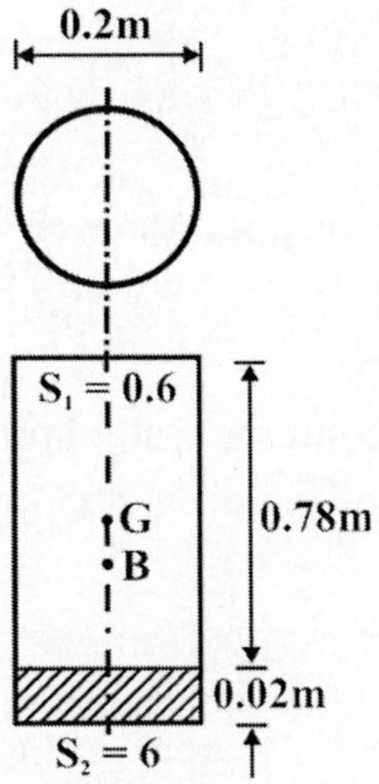

Q. 22 A triangular plate of 1 m base and 1.5 m altitude is immersed in water. The plane of the plate is inclined at 30^0 with free water surface and the base is parallel to and at a depth of 2 m from water surface. Find the total pressure on the plate and the position of the centre of pressure.

2.13.3: Unsolved Problems:

1. In a U tube differential manometer three different liquids are there as shown in the figure. Water is marked with dotted lines, mercury in silver/grey and oil is marked with slanting lines. Assuming that the gage pressure is p kPa 1 =10 kPa, Find the height "d" of the mercury on the unconstrained side.
    i. (Ans: 0.075 m)

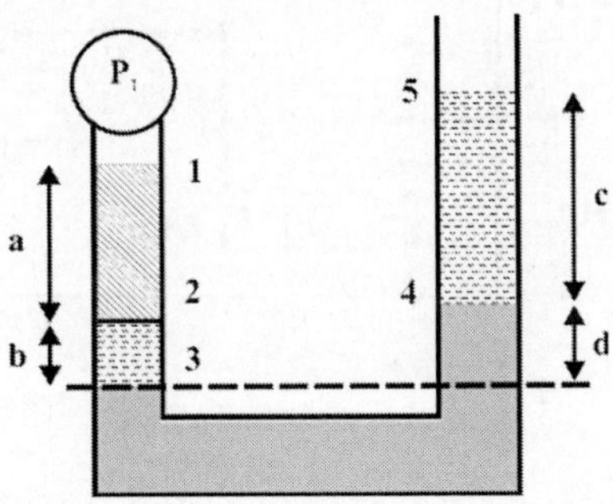

2. A closed tank contains 1.5 m of SAE 30 oil, 1 m of water, 20 cm of mercury, and an air space on top, all at 20°C. If P_{bottom} = 60 kPa, what is the pressure in the air space?
(Hint: Apply the hydrostatic formula down through the three layers of fluid)
(Ans: pair ≈ 10500 Pa)

3. In the given figure, the tank contains water and immiscible oil at 20°C. What is "h" in centimetres if the density of the oil is 898 kg/m3?
(Ans: h = 8 cm)

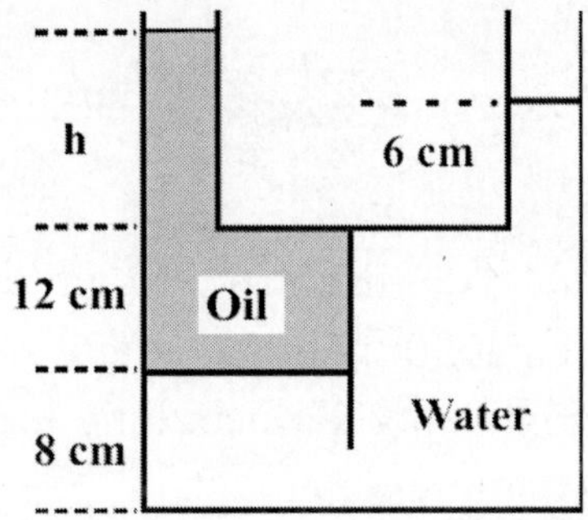

4. In the given figure, water (20°C) and gasoline are open to the atmosphere and are at the same elevation. What is the height h in the third liquid?
(Ans: h = 1.52 m)

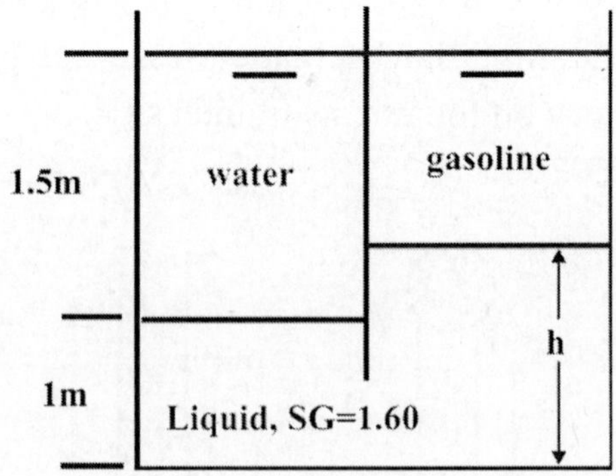

5. In the given figure, both fluids are at 20°C. If surface tension effects are negligible, what is the density of the oil, in kg/m3?

(Ans: h = 1.52 m)

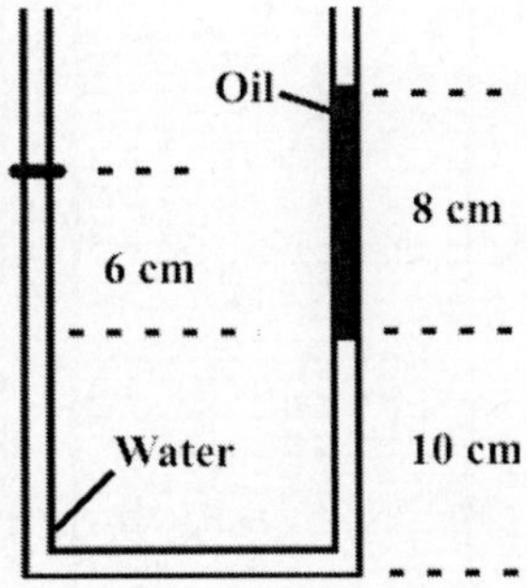

5. A manometer is used to measure the pressure in a tank. The fluid used has a specific gravity of 0.85, and the manometer column height is 55 cm, as shown in Fig. 3–12. If the local atmospheric pressure is 96 kPa, determine the absolute pressure within the tank.
(Ans: P = 100 kPa)

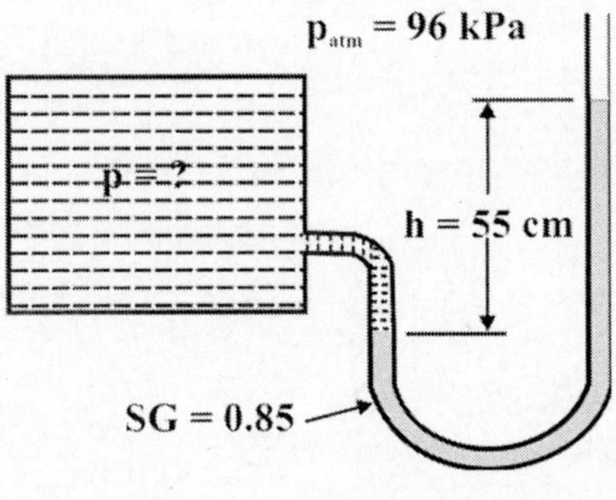

6. The water in a tank is pressurized by air, and the pressure is measured by a multifluid manometer as shown in Figure below. The tank is located on a mountain at an altitude of 1400 m where the atmospheric pressure is 85.6 kPa. Determine the air pressure in the tank if $h_1 = 0.1$ m, $h_2 = 0.2$ m, and $h_3 = 0.35$ m. Take the densities of water, oil, and mercury to be 1000 kg/m^3, 850 kg/m^3, and 13,600 kg/m^3, respectively.
(Ans: P_1 = 130 kPa)

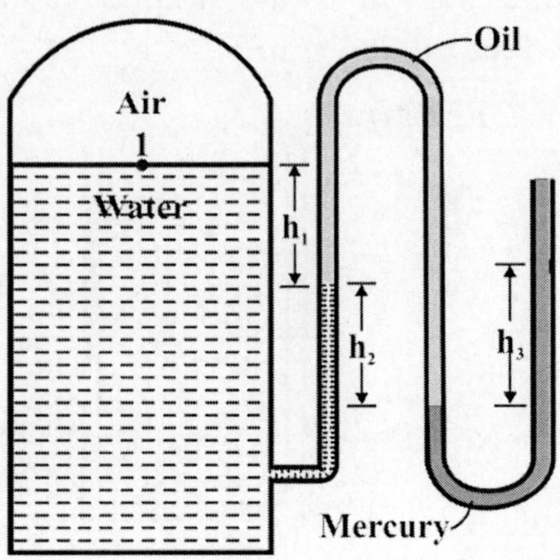
Oil
Air
1
Water
h1
h2
h3
2
Mercury

PRACTICAL: STUDY OF PRESSURE MEASURING DEVICES

Objective:

Study of pressure measuring devices

Theory:

Normal stress on any plane through a fluid element at rest is equal to a unique value called the fluid pressure P. It is expressed in N/m^2 or Pascal as pressure intensity or in terms ofpressure head as 'm'. It can be measured by devices like piezometer, manometer, mechanical gauge or pressure transducer.

Manometer:

Manometer is a device used to measure the fluid pressure by balancing pressure against thecolumn of liquid in static equilibrium. The different types of manometers are listed below.

1. **Simple Manometer**
   - Piezometer
   - Simple U tube manometer
   - Well type U tube manometer
2. **Differential Manometers**
   - Inverted U tube manometer
   - Upright U tube Manometer
   - Micromanometer

1. Simple Manometer:

This measures pressure at a point.

a) **Piezometer:** Piezometer is simple device for measuring pressures of liquids. It consists of a glass tube in which the liquid can rise freely without overflowing. The height of the liquid in the tube above a given datum gives the value of the piezometric head directly. Piezometers measure the pressure above the local atmospheric pressure. It fails to measure vacuum.

b) **Simple U Tube Manometer:** A 'U' tube manometer consists of two tubes joined at one end to form a U – shaped tube. A 'U' tube manometer can be used to measure pressure in any fluid; liquid and gas. Pressure above and below atmospheric can be measured with it. U- tube may used upright , inclined or inverted. Normally the one end of manometer is connected to the gauge point and other is open to the atmosphere. The pressure in the pipe is measured by recording the difference in the level of manometric liquid in two limbs.

c) **Well Type U Tube Manometer:** In the two limb U -tube manometer it is necessary to read the levels of manometric liquid in both tubes to find difference in the levels. In well type manometer a well or basin or reservoir of large cross sectional area is provided on one limb. The fluctuation in the level of the well are very small as compared to the fluctuation in the other limb hence any change in levels of well may be neglected. Fluctuations in the levels of the well may acts as a reference level for other limb. The limb of small cross section area may be upright or inclined depending upon accuracy required. The well side limb is connected to the gauge point and neglecting the change in the level of liquid in the well, the change in the level of liquid in the other limb is noted, which is pressure head at the gauge point.

2 Differential Manometer:

These manometers are used to measure pressure difference between two points.

A) **Upright U Tube Differential Manometer:** It consists of U-tube connected to the points between which pressure difference is to be measured. A heavier manometric liquid occupies lower part of U -tube. These are suitable to measure heavy pressure difference.

B) **Inverted U Tube Manometer:** It consists of inverted u-tube to which an air valve is provided at the top of U -tube. A lighter manometric fluid; liquid or gas occupies upper part of U-tube. These are suitable for the measurement of small pressure difference in liquids.

C) **Macro Manometer:** For measurement of very small pressure difference or pressure measurement with very high precision, micro manometer is used. It consists of u-tube provided with two transparent basin or well of wide section at the top of two limbs. The manometer contains two manometric liquids of different specific gravity and immiscible with each other as well as with the fluid for which the pressure difference is to be measured.

2. BOURDON PRESSURE GAUGE:

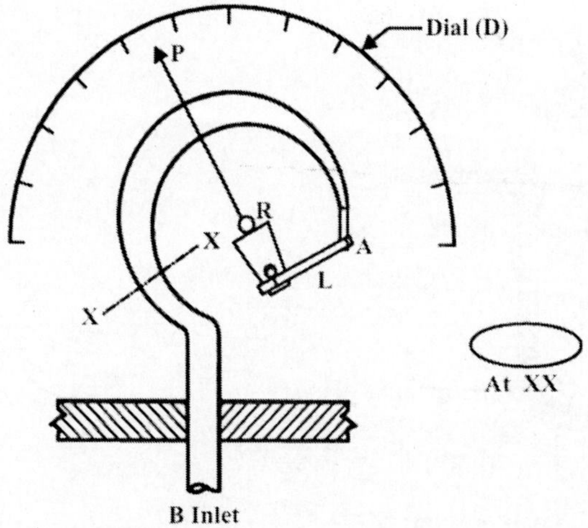

The pressure responsive element in the gauge is a tube of steel or bronze, which is of elliptical cross section and curved in the form of a circular arch. The tube is closed at its outer end of the tube through which the fluid enters is rigidly fixed to frame. When gauge is connected to the gauge point, fluid enters the tube which increases internal pressure making elliptical section circular thus causing tube to straighten out slightly. This outward movement is indicated as pressure on a circular dial through an arrangement of link pinion and indicator. The dial gauge is so calibrated that it reads zero when the pressure inside the tube equals the local atmospheric pressure. When a vacuum gauge is connected to partial vacuum, the tube tends to close there by moving the pointed in anticlockwise direction, indicating the negative or vacuum pressure.

$h_M = S_2h_2 - h_1S_1$

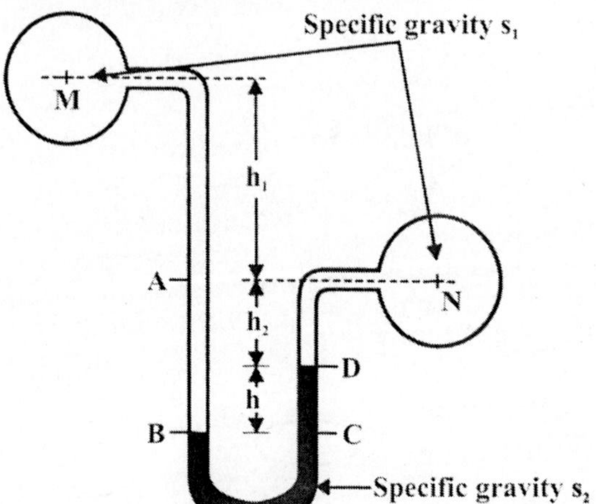

$h_N = -S_2h_2 - S_1h_1$

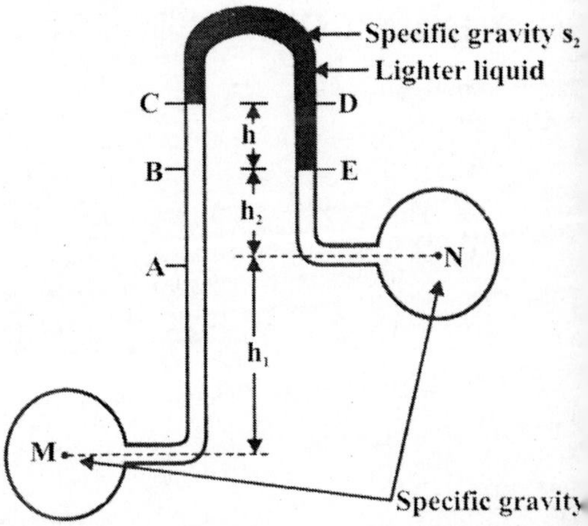

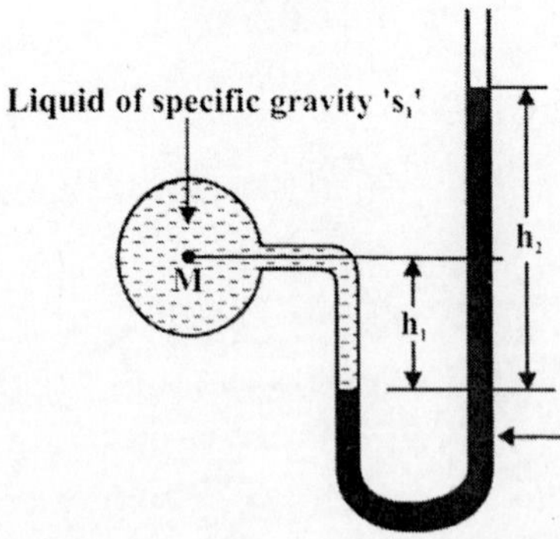

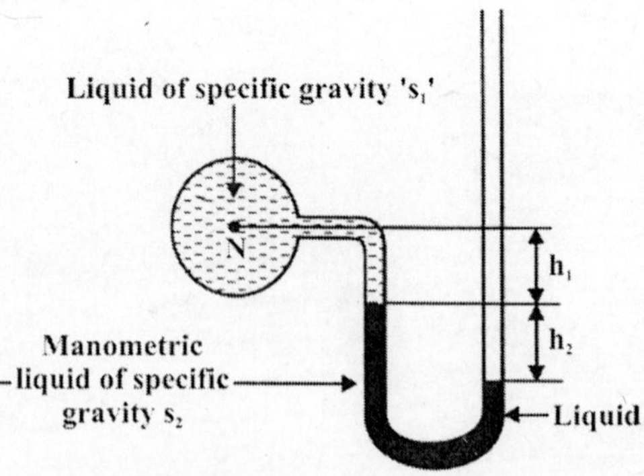

$h_M - h_N = h(S_2 - S_1)$ m of water column

$h_M - h_N = h(S_2 - S_1) + h_1S_1$ m of water column

$$l = \frac{h}{\sin\theta}$$

$$h_M = S_2h_2 - S_1h_1$$

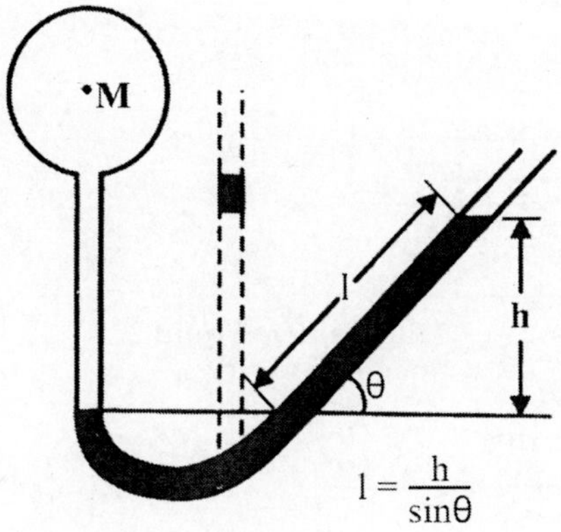

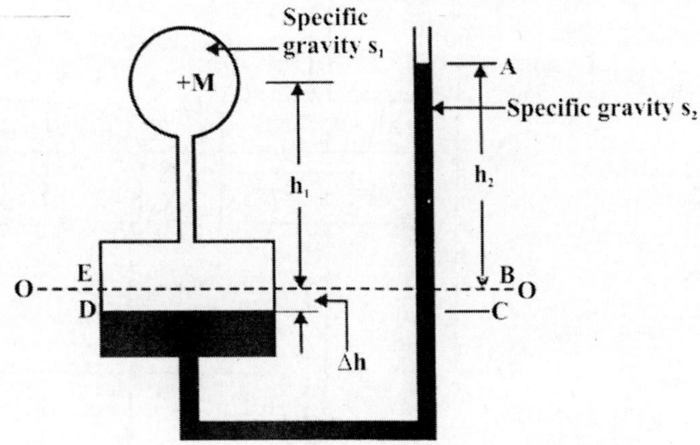

$$h_A - h_B = h\left\{S_3 - S_2\left(1 - \frac{a}{A}\right) - S_1\left(\frac{a}{A}\right)\right\}$$

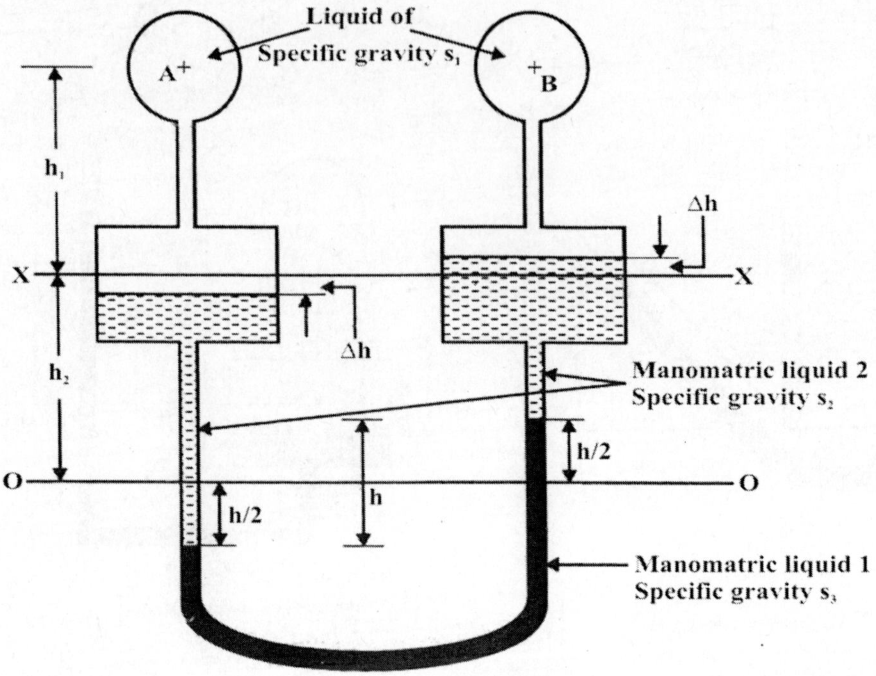

Exercise:

1) Recognize the type of simple manometer and measure the pressure head at a point.

- Type of Simple manometer:
- Pressure at point M =
- Type of Simple manometer:
- Pressure at point N =

2) Recognize the type of differential manometer and measure the pressure difference between points

- Type of Differential manometer:
- Pressure difference h_M - h_N =

3) Measure the pressure / vacuum using both the types of mechanical gauges

4) Name of Gauge:

- Pressure at M :
- Name of Gauge :
- Pressure at N : mm of Hg =

- Pressure Difference =

Conclusion: Student should write the conclusion by his own based upon the above study.

QR CODES FOR SUPPORTING VIDEO LINKS

(1) **(2)** **(3)** **(4)**

REFERENCES AND SUGGESTED READINGS

1. Hydraulics & Fluid Mechanics by Modi and Seth, Standard Book House
2. Theory and Applications of Fluid Mechanics—K. Subramanya- Tata McGraw
3. Fluid Mechanics by R.J. Garde, A.J Mirajgaonkar, SCITECH Publication
4. Fluid Mechanics by Streeter & Wylie, Tata McGraw Hill.

https://archive.nptel.ac.in/noc/courses/noc19/SEM2/noc19-ce28.

3 BUOYANCY AND FLOTATIO

UNIT SPECIFICS

Through this unit we have discussed the following aspects:

- *Concept of buoyancy and Archimedes Principle*
- *Principle of flotation*
- *Stability of submerged and floating bodies*
- *Concept of metacentre and its analytical determination*
- *Experimental determination of metacentre*

This is followed by large number of solved examples. The students are encouraged to solve the objective questions, long answer questions and numerical problems to judge ones understanding. A list of references for additional reading is provided at the end.

RATIONALE

This unit introduces concept of buoyancy which is important is stability analysis of floating and submerged bodies. The metacentre and its role in stability of floating and submerged bodies gives basic information to Civil Engineering students who are further interested in studying Naval Architecture and Ocean Engineering

PRE-REQUISITES

Mathematics: Derivatives (Class XII)

Physics: Mechanics (Class XII)

Fluid Mechanics: Unit I and II

UNIT OUTCOMES

List of outcomes of this unit is as follows:

(At the end of this unit, students will understand..)

U3-O1: Concept of buoyancy and Archimedes Principle

U3-O2: Stability of floating and submerged bodies

U3-O3: Procedure to determine metacentre analytically and experimentally

Unit-3 Outcomes	EXPECTED MAPPING WITH COURSE OUTCOMES (1-WeakCorrelation;2-Mediumcorrelation;3-StrongCorrelation)					
	CO-1	CO-2	CO-3	CO-4	CO-5	CO-6
U3-O1	3	-	1	-	-	-
U3-O2	3	-	1	-	-	-
U3-O3	3	-	-	-	-	-

3.1 Buoyancy:

When a body is immersed partially or fully in a fluid it is acted upon by a vertical force in the upward direction opposite to the self-weight of the body. This force by virtue of which the body 'buoys' or 'floats' is called the buoyant force and its point of application is known as the 'centre of buoyancy'.

It is very clear that if the buoyant force is less than weight of the body, the body will sink otherwise it will float. This is the reason why some solids float and some sink in a liquid.

The wave height measuring instrument 'wave rider buoy' works on the principle of buoyancy. Buoyancy plays a major role in design on ships, submarines, torpedoes. The buoyant force can be calculated either by using laws of hydrostatics or by using the Archimedes Principle.

3.2 Archimedes Principle:

When a body is immersed partially or fully it is lifted up or buoyed up by a force equal to the weight of the fluid displaced by the body.

Consider a body submerged in a fluid of constant mass density as shown in figure 3.1. The body is divided into elementary cylinders like 'AB' of area dA as shown in figure 3.1

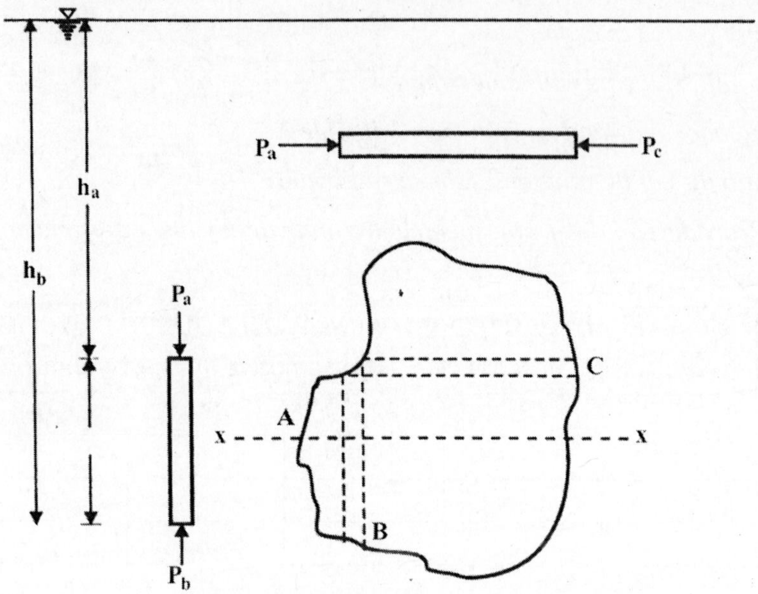

Fig. 3.1 Archimedes principle

Vertically upward force acting on cylinder = $P_b \cdot dA - P_a \cdot dA$

$\therefore$ Total vertically upward force $F = \int (P_b - P_a)\, dA$

But $P_b - P_a = \gamma (h_b - h_a) = \gamma h$

$$\therefore \quad F = \int \gamma h\, dA = \gamma \int h\, dA$$

$$\therefore \quad F = \gamma V \tag{3.1}$$

where V is the volume of fluid displaced by the body and γ V is the weight of the fluid displaced by the body. Thus, buoyant force is equal to weight of the fluid displaced by the body.

If a horizontal cylinder AC is considered, then as A and C are at the same level

$$P_a = P_c \tag{3.2}$$

Horizontal force on cylinder = $(P_a - P_c)\, dA = 0$

Thus, a buoyant force has only vertical component. It is also evident that as the buoyant force is equal to weight of the fluid displaced by the body, it will pass through centre of gravity of displaced fluid. The point of application of buoyant force is known as centre of buoyancy

Thus, a buoyant force is equal to weight of the fluid displaced by the body acting in vertically upward direction and passing through centre of gravity of displaced fluid.

3.3 Principle of Floatation:

Consider a body of weight 'W' completely immersed in a liquid (figure 3.2).

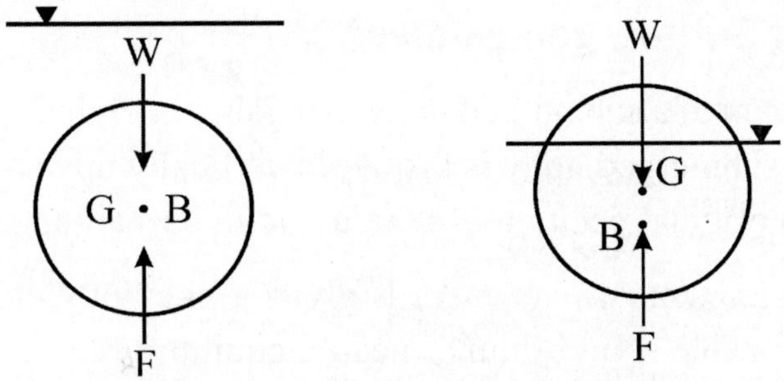

(a) (b)

Fig. 3.2 (a) Principle of floatation

Whether the body will sink or float in the liquid will depend upon weight of the body and buoyant force F i.e specific gravity of the body and specific gravity of the liquid (because specific weight can be determined if specific gravity is known). It is obvious that weight of the body 'W' and buoyant force F should lie along same line to avoid rotation.

Consider specific gravity of body is greater than that of liquid. We know,

Weight of fluid displaced by body = Buoyant force F

∴ Buoyant force F < Weight of the body W and body will sink.

If specific gravity of body is same as that of liquid and the entire body is submerged, the centre of buoyancy B which is centroid of displaced volume of fluid and centre of gravity G will coincide as shown in figure 3.2 and body will remain stable anywhere in the liquid.

If specific gravity of body is less than that of liquid weight of the fluid displaced by the body i.e. buoyant force will be greater than that of weight of fluid and hence body will start rising. The rise will continue till weight of the body and buoyant force attain an equilibrium. That is weight of fluid displaced by body is equal to weight of body. As displaced volume of fluid will be reduced to reduce the weight of fluid displaced by the body to attain equilibrium 'B' will be shifted downwards and body will float.

Thus, in floating condition

Weight of the body = Weight of fluid displaced by the body = Buoyant force

This is known as principle of floatation.

3.4 Stability of Submerged Bodies:

Stability is the tendency of a submerged body to regain its original position when disturbed slightly. If the submerged body is displaced vertically upwards or downwards and if it is gaining its original position back it is said to be stable.

However, if a small angular displacement is given to the body, it will be either in stable equilibrium, unstable equilibrium or neutral equilibrium.

The word equilibrium suggests that weight of the body is balanced by the buoyant force.

However due to angular displacement a rotational effect will lead the body to be in stable, unstable or neutral equilibrium conditions.

3.4.1 Stable Equilibrium:

A body of weight 'w' is completely submerged in a fluid and centre of gravity 'G' lies below centre of buoyance B (figure 3.3 a), If a small angular displacement is given to the body in anticlockwise direction, a restoring clockwise couple will be formed due to change in the alignment of G and B. As a result, the body will regain its original position. This is known as stable equilibrium.

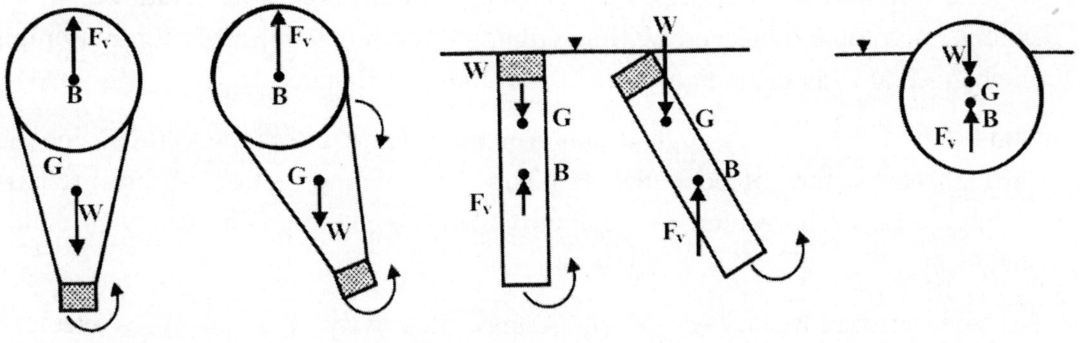

(a) Stable equilibrium (b) Unstable equilibrium (c) Neutral equilibrium

Fig. 3.3 Types of equilibrium

3.4.2 Unstable Equilibrium:

If the centre of buoyancy 'B' lies below the centre of gravity 'G' as shown in figure 3.3 b, the small angular displacement in the anticlockwise direction will be enhanced due to additional anticlockwise moment caused by B and G. As a result, body will never regain its original position and will continue to tilt further. This is unstable equilibrium.

3.4.3 Neutral Equilibrium:

If both G and B coincide, any angular displacement does not shift the position of G and B, as a result body is in neutral equilibrium (figure 3.3c).

3.5 Stability of Floating Bodies:

As discussed in section 3.4 the submerged body is in unstable equilibrium if B is below G. However, the floating body may remain in stable equilibrium even though B is below G. This is due to an additional parameter called 'metacentre' expressed as 'metacentric height'. The position of metacentre plays an important role in stability of floating bodies.

3.5.1 Metacentre and Metacentric Height :

Consider a body floating upright in equilibrium as shown in figure 3.4. The forces acting are weight of the body and buoyant force which are equal and collinear. Due to small angular displacement (angle of heel) θ, centre of buoyance shifts from B to B_1 due to change in volume displaced by fluid (figure 3.4) while the centre of gravity remains unchanged at G as the weight of the body does not change,

If the axis of body passing through B and G is extended it meets the vertical line drawn through new centre of buoyancy B_1. The point of intersection of the line passing through axis of body and vertical through new centre of buoyancy is called as metacentre M.

The distance between metacentre M and Centre of gravity G is known as metacentric height.

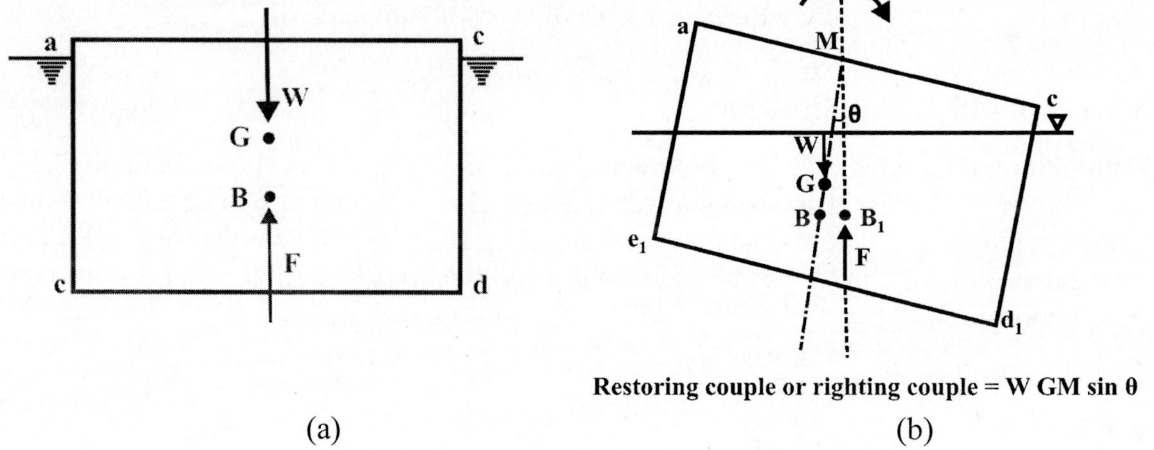

Fig. 3.4 Metacentric height

3.5.2 Stable Equilibrium:

Consider a body floating in a liquid such that its centre of gravity G is above centre of buoyancy B as shown in figure 3.5.

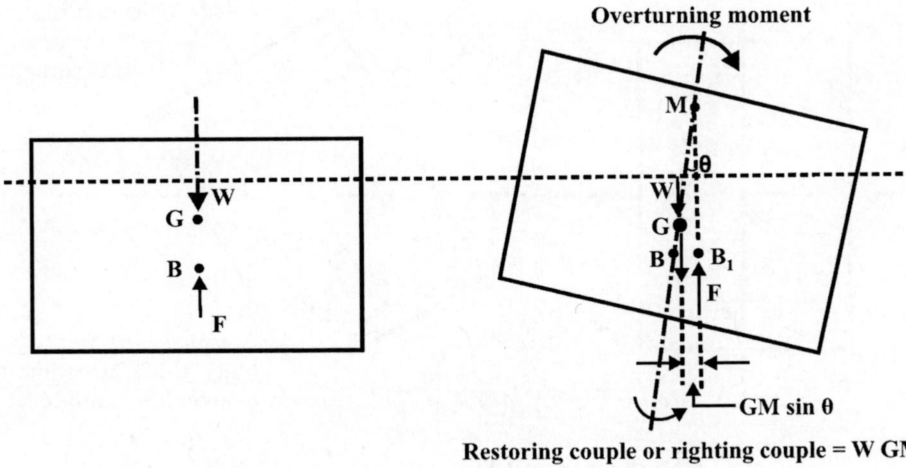

Fig. 3.5 Stable equilibrium (M above G)

If a small angular displacement is given to the body in clockwise direction B will shift B_1 and G will remain in its position. M is the meta centre which lies of above G. The self-weight W and buoyant force F form a restoring couple of magnitude W x GM sin θ in anticlockwise direction due to which body regains its original position.

Thus, if M is above G i.e., GM is positive the floating body is in stable equilibrium.

3.5.3 Unstable Equilibrium:

If the body is floating in such a way that after a small angle of heel, the meta centre lies below centre of gravity, the weight of body W and buoyant force F form the couple in the same sense as that of angular displacement, tilting the body further and the body never comes back to its original position (figure 3.6).

Thus, if M lies below G i.e., GM is negative, the floating body will be in a state of unstable equilibrium. This is the reason why ships are designed in such a way that the engine room, machine room is always on the ground floor in addition to ballast so that 'G' will always remain below M.

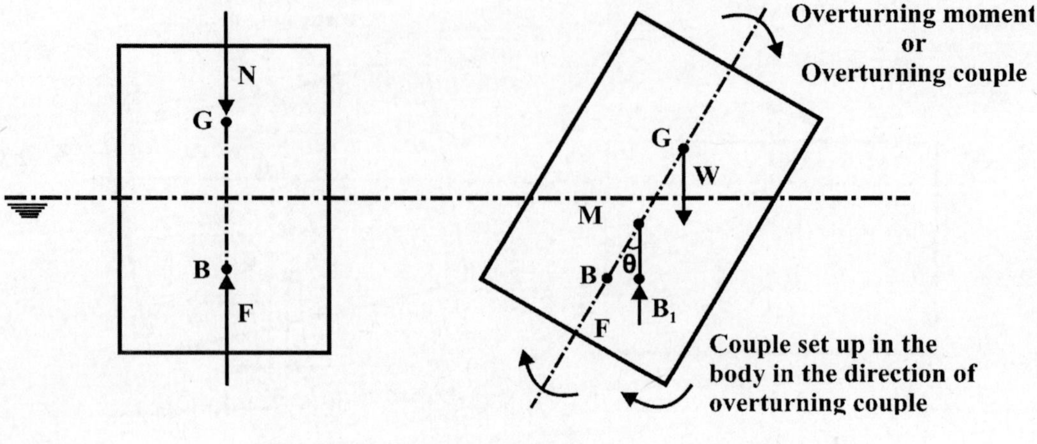

Fig. 3.6 Unstable equilibrium (M below G)

3.5.4 Neutral Equilibrium :

If M and G coincide i.e. GM = 0 the floating body remains in neutral equilibrium.

3.6 Analytical Determination of Metacentric Height :

Figure 3.7 shows plan, section and tilted view of a ship. The centre of gravity is 'G' and centre of buoyancy is B. Due to small angular displacement 'θ', centre of buoyancy shifts from B to B_1, due to change in immersed portion of ship from abcd to a_1 b_1 c_1 d_1.

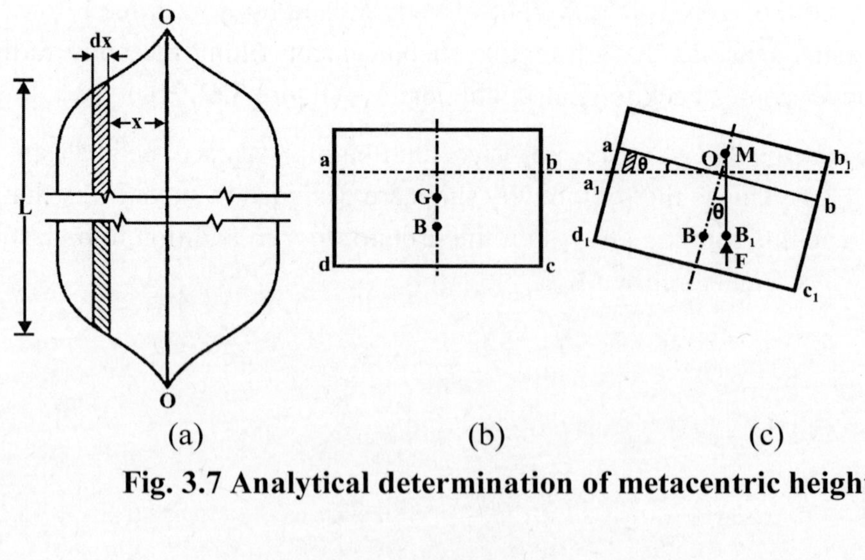

Fig. 3.7 Analytical determination of metacentric height

The anticlockwise moment caused due to shifting of centre of buoyancy from B to B_1 about axis of ship

$= F \cdot B B_1 = F \cdot BM \cdot \tan\theta = F \cdot BM \cdot \theta$ ($\tan\theta \simeq \theta$ for small θ) (3.3)

As weight of the ship will remain same and portion a O a_1 (coming out immersion) is same as b O b_1.

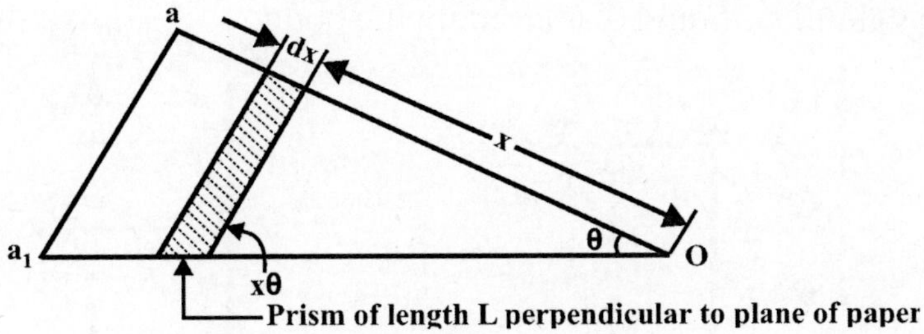

Fig. 3.8 Elementary prism

Consider 2 small elementary prisms of thickness dx at distance 'x' from longitudinal axis of ship (figure 3.8).

Volume of elementary prism = $(x\,\theta) \times dx \times L$

where '$x\theta$' is inclined side of prism and L is length of prism.

Weight of liquid in prism = $\gamma\,(x\,\theta)\,dx\,L$.

∴ Moment of weight of liquid in both prisms about longitudinal axis

$$OO = 2\gamma\,(x\,\theta)\,dx\,L\,x.$$

∴ For the entire wedges, moment

$$M = 2\gamma\,\theta \int x^2\,(L\,dx)$$

$$L\,dx = dA \qquad \text{i.e. area of strip at water line.}$$

∴ $\int x^2 \cdot L\,dx$ = Second moment of area of the prism about OO.

∴ $2\int x^2 \cdot L\,dx$ = Moment of inertia of ship at liquid surface about longitudinal axis

$$OO = I$$

∴ $$M = \gamma\,\theta\,I \qquad (3.4)$$

For equilibrium equating (3.3) and (3.4)

$$F \cdot BM \cdot \theta = \gamma\,\theta\,I$$

$$\therefore \qquad BM = \frac{\gamma I}{F} = \frac{\gamma I}{\gamma V}$$

$$\therefore \qquad BM = \frac{I}{V} \qquad (3.5)$$

where 'V' the volume of liquid displaced by the body.

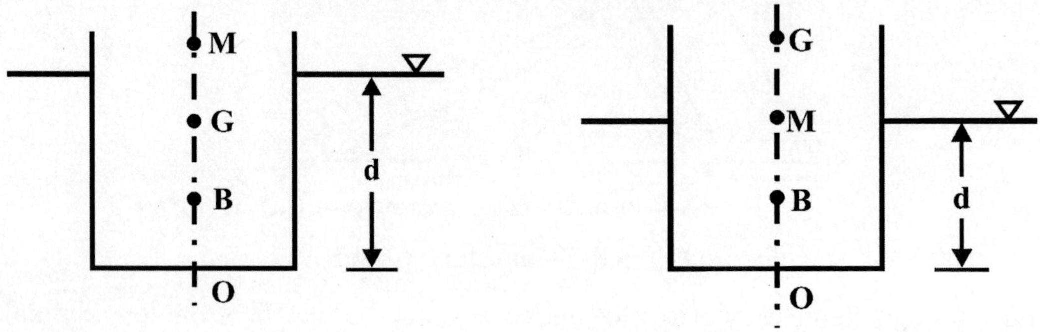

Fig. 3.9 The stable and unstable equilibrium due to relative positions of M and G

$OB = \frac{d}{2}$ = depth of immersion	$OB = \frac{d}{2}$ = depth of immersion
BG = OG – OB	BG = OG – OB
GM = BM – BG = positive	GM = BM – BG = negative
i.e. $BM > BG$	i.e. $BM < BG$

Thus if $BM > BG$, GM is positive and body will float in stable equilibrium and vice-versa.

3.7 Experimental Determination of Metacentric Height :

A small jockey weight 'm' is initially kept at the centre of deck of the ship of weight W. When the jockey weight is displaced by distance 'x', ship tilts and C.G. and centre of buoyancy are shifted to new position G_1, and B_1 respectively (as the jockey weight changes its position the CG also changes it position from G to G1 unlike the previous cases). The small angular displacement (angle of heel, θ) is measured with the help of plumb bob moving over graduated scale. The distance 'y' moved by plumb bob of length '*l*' over the graduated scale yields the angle of heel as,

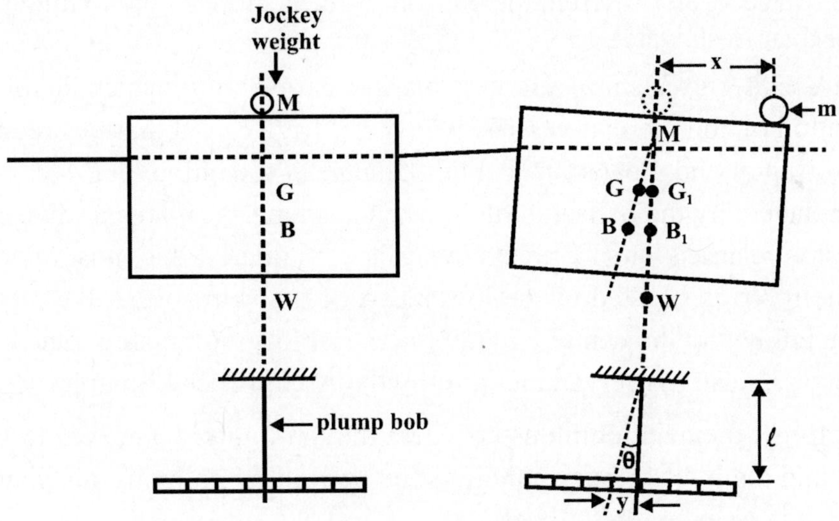

Fig. 3.10 Experimental Determination of Metacentre

$$\tan\theta = \frac{y}{l} \tag{3.6}$$

Disturbing moment causing by moving the jockey weight = m . x

Restoring moment induced by shifting W from G to G_1 = W . GM . sin θ

= W . GM . tan θ (θ is small)

For equilibrium,

$$W \cdot GM \cdot \tan\theta = m \cdot x$$

$$GM = \frac{m \cdot x}{W \tan\theta} = \frac{m \cdot x}{w \cdot \frac{y}{l}}$$

Thus, metacentre can be determined experimentally.

To elaborate buoyancy in most practical approach to civil engineering students, few practical as well as industrial applications are discussed below,

3.8 Industrial (practical) applications of Buoyancy:

Buoyancy and flotation have many practical applications ranging from sports mechanics to shipping industry to instrumentation. It is a well-known fact that all sports related to swimming such as swimming, diving, water polo require knowledge about buoyancy and flotation which is present naturally in fish. It also comes handy for surfers in sea water wherein the buoyant force is more owing to higher density of sea water than fresh water.

The naval architecture, ship building, marine engineering which involve design and operation of commercial vessels, war ships, frigates, submarines require knowledge of buoyancy and floatation to both designers and pilots and sea men. One must understand why the ballast is filled when a submarine wants to dive deep in sea and why it is released when it wants to surface. Similarly, one must understand that the engine room is situated in the lower part of the ship so that the metacentre would never fall below the centre of gravity for stability of the ship. One can say that the shipping industry is served the most by the Archimedes Principle.

Let us not forget the aviation industry where though lighter compared to water, 'air' is a fluid and the laws of buoyancy must be obeyed by balloons, air planes, helicopters which comes naturally to 'birds'.

The instruments such as wave rider buoy which measures the wave height or hygrometer (used in measurement of density) or lactometer (used for measurement of purity of milk called as "degree"). The inverter which is a source of uninterrupted power supply for domestic as well as industrial use has "the distilled water" the level of which can easily be understood by using a "float" kind of indicator. The rain gauges having a floating mechanism installed in it were very common before the advent of automatic rain gauges and digital measurements. The principle of buoyancy was used for design of automatic gates by Bharatranta M. Vishwesaraiya for while designing Bhatghar dam in Pune district of Maharashtra commissioned in 1927 are still operative. All the above examples prove that the phenomenon of buoyancy and floatation must be understood thoroughly who are in any kind of business related to "fluids".

3.9 Solved Examples:

Ex. 3.1 : A cube of sides 'a' floats in water. Compute ranges of specific gravity of cube material so that cube will float with axis vertical. Will the range change for other liquids? Justify.

Solution :

Weight of liquid displaced = weight of body

$$\text{Volume x } \gamma = S\,\gamma\, a^3$$

$$\therefore \quad \text{Volume of liquid} = S\, a^3$$

$$\therefore \quad \text{depth of immersion } h = \frac{S\,a^3}{a^2}$$

$$h = Sa$$

$$OG = \frac{d}{2} = \frac{a}{2}$$

$$OB = \frac{Sa}{2}$$

$$\therefore \quad BG = \frac{a}{2}(1 - S)$$

$$BM = \frac{I}{V} = \frac{a^4}{12\,S\,a^3} = \frac{a}{12\,S}$$

For stable equilibrium

$$BM > BG$$

$$\frac{a}{12\,S} > \frac{a}{2}\,(1 - S)$$

$$\frac{1}{6} > S\,(1 - S)$$

$$S^2 - S + \frac{1}{6} > 0$$

$$(S - 0.789)\,(S - 0.211) > 0$$

$$\therefore \quad S > 0.789 \text{ or } S < 0.211$$

Ex. 3.2: A cylindrical buoy is 2 m in diameter and 2.5 m long and weights 22 kN. The specific weight of sea water is 10.25 kN/m^3 show that buoy does not float with its axis vertical

Solution:

$$\text{Weight of liquid displaced} = \text{Weight of solid}$$

$$\text{Volume of liquid displaced x specific weight} = 22$$

$$\therefore \quad \text{Volume of liquid displaced} = \frac{22}{10.25} = 2.15 \text{ m}^3$$

$$\therefore \quad \text{depth of immersion} = \frac{2.15}{Area} = \frac{2.15}{\frac{\text{Л}}{4} \text{x} 4} = 0.68 \text{ 'm'}$$

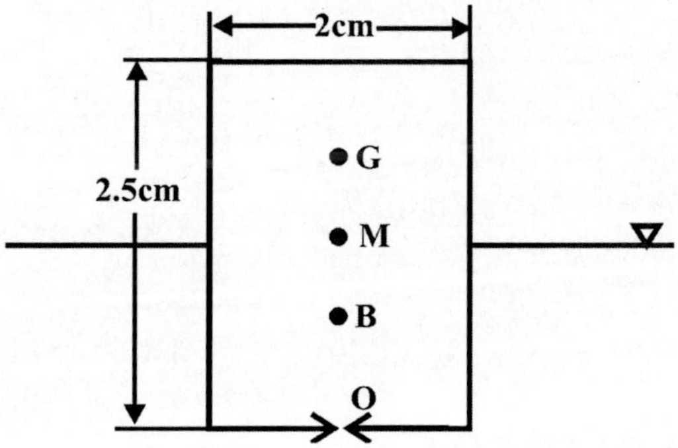

Fig. Ex. 3.2

$$\therefore \quad OB = 0.34 \text{ m}$$

$$OG = 1.25 \text{ m}$$

$$BG = 1.25 - 0.34 = 0.91$$

$$BM = \frac{I}{V} = \frac{\frac{\text{Л}}{64} \text{x} 24}{2.15} = 0.36$$

$$BM < BG \therefore GM \text{ is negative}$$

$\therefore$ buoy does not float with axis vertical.

Ex. 3.3:	Find the percent volume of an ice berg above the water surface if it floats in sea water. Assume density of sea water 1010 kg/m^3 and density of ice berg 920 kg/m^3.

Solution :

$$\gamma = \rho g$$

Weight of water displaced = Weight of ice berg

∴ Volume of water displaced x Sp. Wt. of water
= Vol. of ice berg x Sp. Wt. of ice berg

∴ Volume of water displaced x ρ_W x g
= Vol. of ice berg x ρ_{ice} x g

$$\text{Volume of water displaced} = \text{Volume of ice berg} \times \frac{920}{1010}$$

∴ Volume of water displaced = 0.9109 volume of ice berg

$$A_{ice} \times h = 0.9109 \times A_{ice} \times H$$

(water surface area and ice area will be same)

∴ $$h = 0.91\ H$$

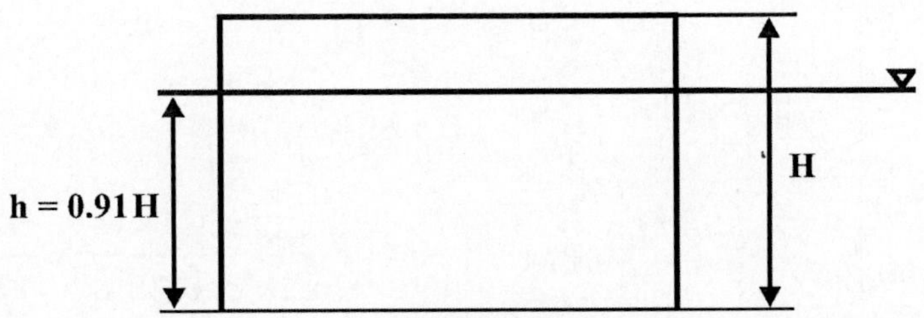

Fig. Ex. 3.3

∴ 8.91% by volume of an ice berg above water surface.

Ex. 3.4 :	A wooden cylinder of specific gravity 0.66 is required to float in oil of specific gravity 0.88. If diameter of cylinder is D and length L, calculate the limiting ratio between L and D for cylinder to float in stable equilibrium vertically.

Solution :

Weight of oil displaced = Weight of cone

$$\left(\frac{\pi}{4} D^2 \times h\right) \times 0.88\, \gamma_W = \left(\frac{\pi}{4} D^2 \times L\right) \times 0.66\, \gamma_W$$

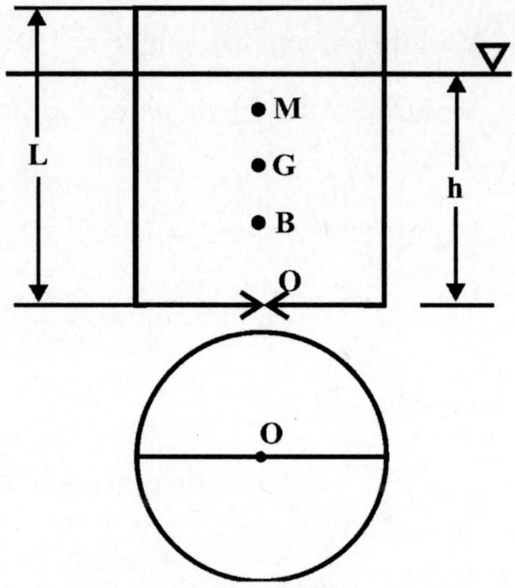

Fig. Ex. 3.4

$$OG = \frac{L}{2}$$

$$OB = \frac{0.75}{2} L$$

$$BG = 0.125\ L$$

$$BM = \frac{I}{V} = \frac{\frac{\pi}{64} \times D^4}{\frac{\pi}{4} \times D^2 \times 0.75\, L} = \frac{D^2}{12\, L}$$

For stable equilibrium

$$BM = BG$$

$$\frac{D^2}{12\, L} = \frac{L}{8}$$

$$\therefore \quad \frac{D}{L} = 1.225$$

Ex. 3.5 : A rectangular block 50 cm long, 25 cm wide and 18 cm deep is floating in a liquid. The shortest axis of block is vertical and the depth of immersion is 15 cm. Calculate the metacentric height and comment on stability of the block. What will happen if longest axis is vertical?

Solution:

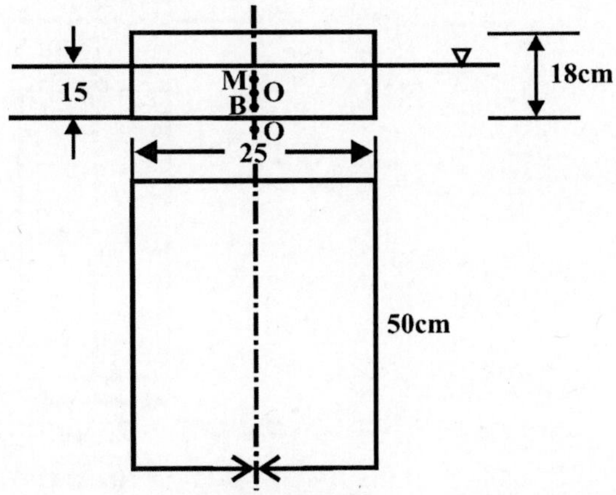

Fig. Ex. 3.5

OG = 0.09 OB = 0.075 BG = 0.015

$$BM = \frac{I}{V} = \frac{0.5 \times (0.25)^3}{12 \times 0.25 \times 0.5 \times 0.15} = 0.035$$

$$GM = 0.0197 \text{ m}$$

$$= 1.97 \text{ cm} \quad \therefore \text{stable}$$

If longer axis vertical

OG = 0.25 OB = 0.075 BG = 0.175

$$BM = \frac{0.18 \times 0.25^3}{12 \times 0.25 \times 0.15 \times 0.18} = 0.035$$

$$GM = 0.035 - 0.175 = -0.14 \text{ unstable}$$

Ex. 3.6 : A wooden block 50 cm long, 25 cm wide and 18 cm deep has its shorter axis vertical with the depth of immersion 15 cm. Calculate the position of metacentre and comment on the stability of the block.

Solution :

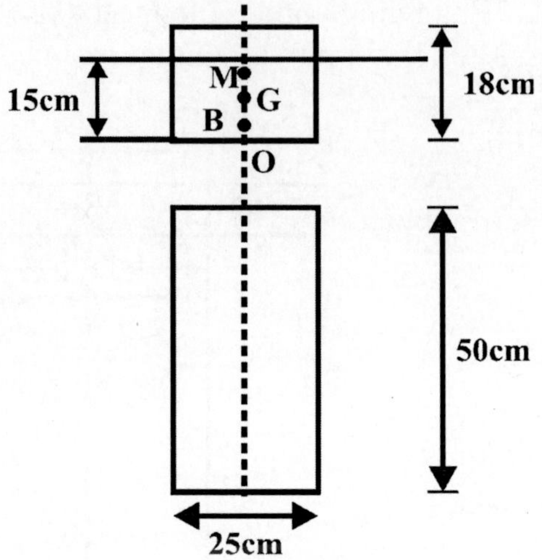

Fig. Ex. 3.6

$$d = 0.15 \text{ m}$$

$$OB = 0.075 \text{ m} \qquad OG = 0.09 \text{ m}$$

$$BG = 0.09 - 0.075 = 0.015 \text{ m}$$

$$I = \frac{0.50 \, x \, 0.25^3}{12} = 6.51 \text{ x } 10^{-4}$$

$$BM = \frac{I}{V} = \frac{6.51 \, x \, 10^{-4}}{0.5 \, x \, 0.25 \, x \, 0.15} = 0.0347$$

$$GM = BM - BG = 0.0197 \text{ positive}$$

∴ block is stable

Ex. 3.7: A cylinder of 20 cm in diameter has a metal base of 3 cm thick whose specific gravity in 8. The upper part of the cylinder is made of a material whose specific gravity is 0.5. Find the maximum height of the composite cylinder for stable equilibrium in the water.

Solution:

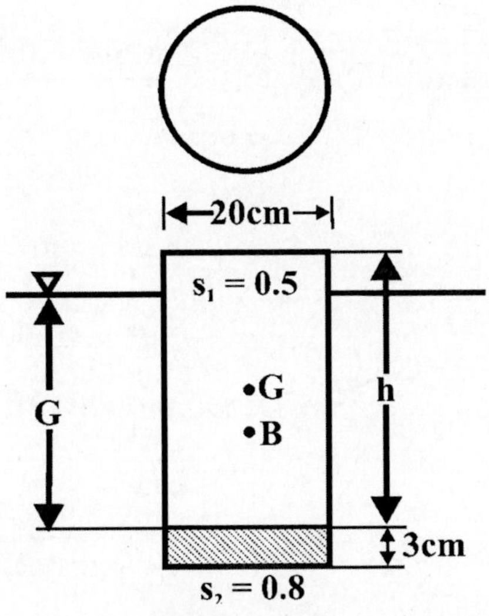

Fig. Ex. 3.7

Distance of centre of gravity of combined solid cylinder from base :

$$OG = \frac{w_1 y_1 + w_2 y_2}{w_1 + w_2}$$

$$OG = \frac{\frac{Л}{4} \times D^2 \times h \times 0.5 \times 9810 \times \left(3 + \frac{h}{2}\right) + \frac{Л}{4} \times D^2 \times 0.03 \times 8 \times 9810 \times 0.015}{\frac{Л}{4} \times D^2 \times h \times 0.5 \times 9810 + \frac{Л}{4} \times D^2 \times 0.03 \times 8 \times 9810}$$

$$OG = \frac{0.5\, h \left(3 + \frac{h}{2}\right) + 3.60 \times 10^{-3}}{0.5\, h + 0.24}$$

Let 'd' be depth of immersion :

Weight of cylinder = Weight of water displaced

$$\frac{Л}{4} \times (0.2)^2 \times h \times 0.5 \times 9810$$

$$= \frac{Л}{4} \times (0.2)^2 \times d \times 9810 + \frac{Л}{4} \times (0.2)^2 \times 0.03 \times 8 \times 9810$$

	$0.5\ h + 0.24 = d$ $OB = \frac{d}{2} = 0.25\ h + 0.12$ $\therefore \quad BG = OG - OB = \frac{1.5\ h + 0.25\ h^2 + 3.6 \times 10^{-3}}{0.5\ h + 0.24} - 0.25\ h - 0.12$ $BM = \frac{I}{V}$ $I = \frac{Л}{4} \times D^4 = \frac{Л}{4} \times (0.2)^4$ $\frac{Л}{4} \times D^2 \times d = \frac{Л}{4} \times (0.2)^2 \times (0.5\ h + 0.24)$ $\therefore \quad \frac{I}{V} = \frac{\frac{Л}{4} \times (0.2)^4}{\frac{Л}{4} \times (0.2)^2 \times (0.5\ h + 0.24)} = \frac{0.04}{0.5\ h + 0.24}$ For stable equilibrium : $M = BG$ $\frac{0.04}{0.5\ h + 0.24} = \frac{1.5\ h + 0.25\ h^2 + 0.0036}{0.5\ h + 0.24} - 0.25\ h - 0.12$ $\therefore \quad 0.04 = 1.5\ h + 0.25\ h^2 + 0.0036 - 0.125\ h^2 - 0.06\ h - 0.06\ h - 0.0288$ $\therefore \quad 0.125\ h^2 + 1.38\ h - 0.0652 = 0$ $\therefore \quad h = \frac{-1.38 + \sqrt{(1.38)^2 + 4 \times 0.125 \times 0.0652}}{2 \times 0.125} = 0.047\ m$ $\therefore$ Height of cylinder = 4.7 + 3 = 7.7 cm

Ex. 3.8:	A cylindrical buoy is 2 m in diameter and 2.5 m long and weights 21.6 kN. The density of sea water is 10055 N/m^3. Show that the buoy does not float with its axis vertical. What minimum pull should be applied to a chain attached to the centre of the base to keep it vertical?

Solution :

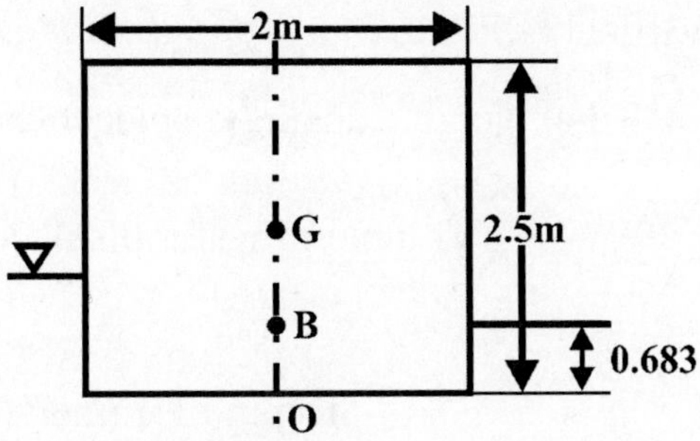

Fig. Ex. 3.8

Volume of liquid displaced :

$$V = \frac{W}{\text{specific weight of liquid}}$$

$$= \frac{21.6}{10.055} = 2.148 \text{ m}^3$$

Depth of immersion :

$$d = \frac{\text{Volume}}{\text{Area in plan}}$$

$$= \frac{2.148}{\frac{\pi}{4} \times D^2} = \frac{2.148}{\frac{\pi}{4} \times 2^2}$$

$$= 0.683 \text{ m}$$

$\therefore$ $OB = \frac{d}{2} = 0.341$ m, OG = 1.25 m

Now, $BG = OG - OB = 1.25 - 0.341 = 0.909$ m

Now, $BM = \frac{I}{V} = \frac{\pi}{64} \times \frac{2^4}{2.148} = 0.3656$ m

As BM < BG, M is below CG.

i.e. GM is negative.

$\therefore$ The cylinder will not float with its axis vertical.

II) Additional weight :

Let an additional pull of 'T' kN be applied at the base to keep the buoy vertical:

∴ Total weight causing displacement W' = W + T

$$\text{Volume of water displaced } V' = \frac{W+T}{10.055}$$

$$\text{Depth of immersion } d' = \frac{V'}{Area\ in\ plan} = \frac{\frac{W+T}{10.055}}{\frac{\pi}{4} \times 2^2}$$

$$OB' = \frac{d}{2} = \frac{W+T}{31.58 \times 2} = \frac{W+T}{63.17}$$

As the additional weight is increased, the combined specific gravity G' starts moving towards new metacentre M' and finally they both coincide and the cylinder treats in neutral equilibrium.

Combined weight (W + T) and buoyant force (W + T) acting upwards is shown in Fig. Ex. 3.42(a).

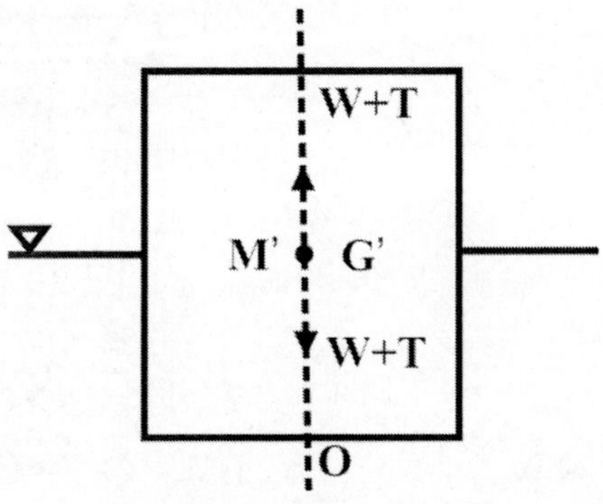

Fig. Ex. 3.8

$$W \cdot OG = (W + T)\, OG'$$

$$\therefore \quad OG' = \frac{W.OG}{W+T} = \frac{W \times 1.25}{W+T}$$

$$B'G' = OG' - OB'$$

$$= \frac{1.25\, W}{W+T} - \frac{W+T}{63.17}$$

	$$B'M' = \frac{I'}{V'} = \frac{\pi}{64} \times 2^4 \times \frac{10.055}{W+T} = \frac{7.897}{W+T}$$ For neutral equilibrium $$B'M' = B'G'$$ $$\frac{7.897}{W+T} = \frac{1.25\,W}{W+T} - \frac{W+T}{63.17}$$ $\therefore$ $$\frac{7.897}{W+T} = \frac{27}{W+T} - \frac{W+T}{63.17}$$ $\therefore$ $$\frac{7.897}{W+T} = \frac{1705.59 - (W+T)^2}{63.17\,(W+T)}$$ $\therefore$ $$498.85 = 170.59 - (W+T)^2$$ $\therefore$ $$(W+T)^2 = 1206.73$$ $\therefore$ $$W+T = 34.738$$ $\therefore$ $$T = 34.738 - 21.6$$ $\therefore$ $$T = 13.13 \text{ kN}$$

Ex. 3.9 :	A solid cone made of a material of 0.8 specific gravity floats in water with its apex downwards. Determine the least open angle if the cone is in stable equilibrium.

Solution :

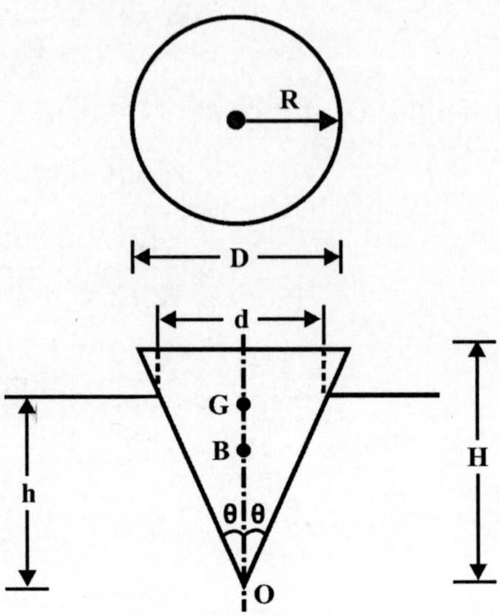

Fig. Ex. 3.9

$$\text{Weight of cone} = 0.8 \times 9810 \times \frac{1}{3} \times \pi R^2 H$$

Let 'h' be depth of immersion.

$$r = \frac{d}{2} = h \tan\theta$$

and $$R = H \tan\theta$$

Weight of liquid displaced = Specific weight of liquid x Volume of liquid displaced.

$$= 9810 \times \frac{1}{3} \times \pi^2 \times h$$

$$= 9810 \times \frac{1}{3} \times \pi \times h^3 \tan^2\theta$$

Weight of solid = Weight of liquid displaced

$$0.8 \times 9810 \times \frac{1}{3} \pi R^2 H = 9810 \times \frac{1}{3} \times \pi \times h^3 \tan^2\theta$$

$\therefore$ $$0.8 \times H^3 \tan^2\theta = h^3 \tan^2\theta$$

$$\therefore \quad \left(\frac{H}{h}\right)^3 = \frac{1}{0.8} \text{ where 'h' is depth of immersion.}$$

$$OB = \frac{3h}{4} \qquad OG = \frac{3H}{4}$$

$$\therefore \quad BG = OG - OB$$

$$\therefore \quad BG = \frac{3}{4}(H - h)$$

$$BM = \frac{I}{V}$$

$$I = \frac{\text{л}}{64} d^4$$

(Moment of inertia of circular section in plane about water surface. Diameter of cone at water surface = d)

$$\text{Volume of cone in water} = \frac{1}{3} \text{л} \, r^2 h$$

$$= \frac{1}{3} \text{л} \times h^2 \tan^2 \theta \times h = \frac{1}{3} \text{л} h^3 \tan^2\theta$$

$$\therefore \quad BM = \frac{\frac{\text{л}}{64} d^4}{\frac{1}{3} \text{л} \, h^3 tan^2\theta} = \frac{\frac{\text{л}}{64} \times 16 \times r^4}{\frac{1}{3} \text{л} h^3 tan^2\theta}$$

$$\therefore \quad BM = \frac{\frac{1}{4} \times h^4 \; tan^4 \theta}{\frac{1}{3} \times h^3 \; tan^2 \theta}$$

$$BM = \frac{3}{4} h \tan^2 \theta$$

For stable equilibrium BM = BG :

$$\therefore \quad \frac{3}{4} h (\tan^2\theta) = \frac{3}{4}(H - h)$$

$$\therefore \quad h \tan^2\theta = H - h$$

$$\therefore \quad h(1 + \tan^2\theta) = H$$

$$\therefore \quad 1 + \tan^2\theta = \frac{H}{h}$$

$$\therefore \quad 1 + \tan^2\theta = \sqrt[3]{\frac{1}{0.8}}$$

$$\therefore \quad \tan^2\theta = 0.0772$$

$$\therefore \quad \tan\theta = 0.27788$$

$$\therefore \quad \theta = 15.52^0$$

$$\therefore \quad \text{apex angle of cone} = 2\theta = 31.05$$

Ex. 3.10 : An equilateral triangular prism having axis 0.8 m long and sides of base 0.5 m floats in water with axis hz and has specific gravity of material 0.8. Calculate water force exerted on vertical and inclined faces of the prism. Also calculate CP in each case w.r.f. water surface.

Solution :

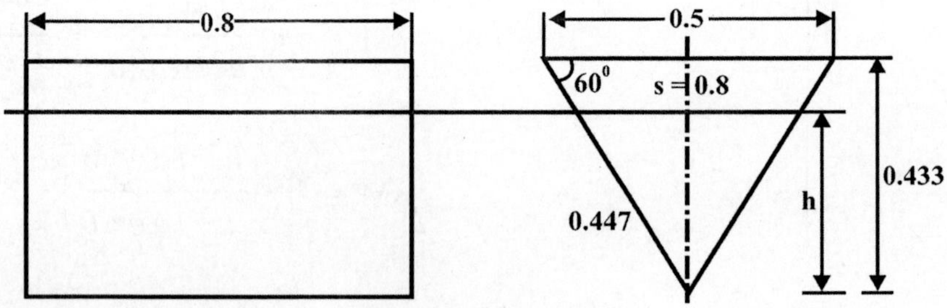

Fig. Ex. 3.10

Weight of liquid displaced = Weight of body

$$\text{Volume of liquid} \times 9.81 = \frac{1}{2} \times 0.5 \times 0.433 \times 0.8 \times 0.8 \times 9.81$$

$$\text{Volume of liquid} = 0.06928 = \frac{1}{2} \times h \times \frac{2h}{\sqrt{3}} \times 0.8$$

	$= \dfrac{0.06928}{\frac{1}{2} \times h \times \frac{2h}{\sqrt{3}} \times 0.8}$ Width of prism at the level of immersion = 0.45 m (using similar triangles) Total pressure on vertical face. $P_I = 9.81 \times \frac{1}{2} \times 0.39 \times 0.45 \times 0.13 = 0.11$ kN Centre of pressure $\bar{h}_1 = 0.13 + \dfrac{\frac{0.45 \times 0.39^3}{36}}{\frac{1}{2} \times 0.39 \times 0.45 \times 0.13}$ $\bar{h}_1 = 0.195$ 'm' from top Inclination of prism $= 60^0$ $\therefore \sin^2\theta = 0.75$ Total pressure on inclined face, $P_2 = 9.81 \times A \times \bar{x} = 9.81 \times 0.447 \times 0.8 \times \frac{0.39}{2}$ $P_2 = 0.68$ kN Centre of pressure $\bar{h}_2 = 0.195 + \dfrac{\frac{(0.8 \times 0.447)^3}{12} \times 0.75}{0.8 \times 0.447 \times 0.195}$ $\bar{h}_2 = 0.259$ 'm' from top.

Ex. 3.11:	A weight 100×10^3 N is moved through a distance of 6 m across the deck of a vessel of 10×10^6 N is moved through a distance of 6 m across the deck of a vessel of 10×10^6N displacement floating in water. This makes a pendulum of 2.5 m swing through a distance 12.5 cm horizontally. Calculate metacentric height of the vessel. **Solution:**

$$\tan\theta = \frac{12.5}{250} = 0.05$$

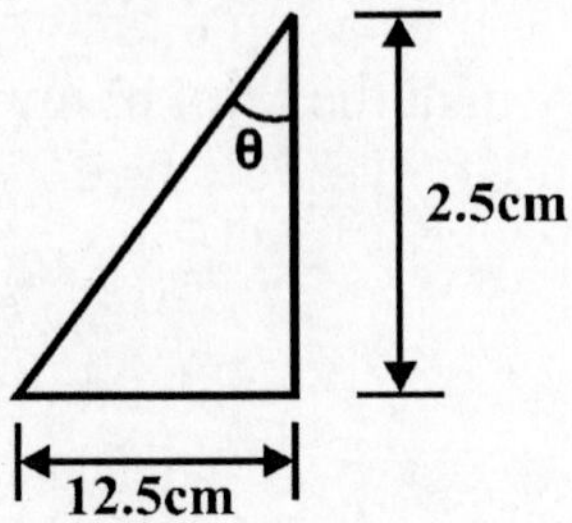

Fig. Ex. 3.11

$$G.M. = \frac{m \, . \, x}{w \tan\theta} = \frac{100 \times 10^3 \times 6}{10 \times 10^6 \times 0.05}$$

where
m = jockey weight
x = displacement of jockey
w = weight of ship
G.M. = Metacentric height

G.M. = 1.2 m

Ex. 3.12 : A wooden cylinder of diameter 'd' and length '2d' floats in water with its axis vertical. Is the equilibrium stable? Take specific gravity of wood = 0.6.

Solution :

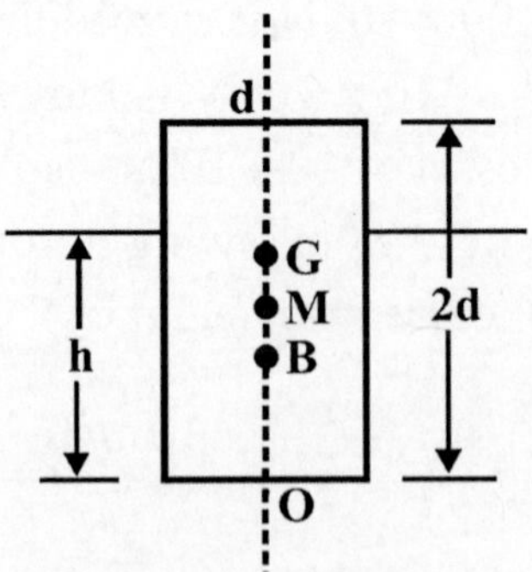

Fig. Ex. 3.12

	Let 'h' be depth of immersion Weight of liquid displaced = weight of the solid ∴Volume of liquid displaced x specific weight of liquid = weight of the solid ∴ $\left(\frac{\pi}{4}\ d^2 \times h\right) \times 9.81 = \frac{\pi}{4} \times d^2 \times 2d \times 0.6 \times 9.81$ ∴ $h = 1.2\ d$ $OG = d$ $OB = \frac{h}{2} = 0.6\ d$ $BG = OG - OB = 0.4\ d$ $BM = \frac{I}{V} = \frac{\frac{\pi}{64} \times d^3}{\frac{\pi}{4} \times d^2 \times 1.2\ d} = \frac{\pi}{64} d^3 \times \frac{4}{\pi d^3 \times 1.2}$ $BM = 0.052$ $BM < BG$ i.e. GM is negative ∴ equilibrium is unstable.

Ex. 3.13 :	A cube of side 'b' floats with one of the its axes vertical in the liquid of specific gravity S_L. If the specific gravity of the cube material is S_C, find ratio of S_L / S_C for the metal centric height to be zero. **Solution :** 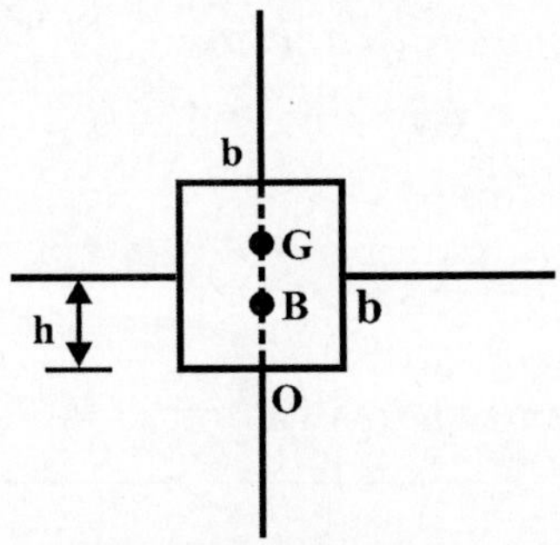 **Fig. Ex. 3.13**

Let 'h' be depth of immersion.

Weight of liquid displaced = weight of solid

$$b \times b \times h \times S_L \times \gamma_W = b \times b \times b \times S_C \times \gamma_W$$

$$\therefore \quad h = b\,\frac{S_C}{S_L}$$

$$\text{i.e.} \quad h = b\,.\,x \left(x = \frac{S_C}{S_L}\right)$$

$$OB = \frac{bx}{2}$$

$$OG = \frac{b}{2}$$

$$BG = OG - OB = \frac{b}{2}(1 - x)$$

$$BM = \frac{I}{V} = \frac{\frac{b^4}{12}}{b \times b \times h} = \frac{b^2}{12h}$$

$$\therefore \quad BM = \frac{b^2}{12 \times b\,.\,x} = \frac{b}{12x}$$

For metacentric height to be zero

$$GM = 0 = BM - BG$$

$$\therefore \quad BM = BG$$

$$\therefore \quad \frac{b}{12} = \frac{b}{2}(1 - x)$$

$$\therefore \quad \frac{1}{6x} = 1 - x$$

$$\therefore \quad 6x^2 - 6x + 1 = 0$$

$$\therefore \quad x = \frac{S_C}{S_L} = 0.211 \text{ or } 0.789$$

$$\therefore \text{ range of values for } \frac{S_S}{S_C} = 4.74 \text{ to } 1.27$$

UNIT SUMMARY

3.10 Exercise :

3.10.1 Objective Questions:

1. When a body floats in water the buoyancy force equals _____

   (a) The weight of volume of water displaced

   (b) The weight of volume of water which is not displaced

   (c) The weight of volume of total water in which the body floats

   (d) None of the above

   Ans: a

2. Buoyant force is ---

   (a) Resultant of up thrust and gravity forces acting on the body

   (b) Resultant force on the body due to the fluid surrounding it

   (c) Resultant of static weight of body and dynamic thrust of fluid'

   (d) Equal to the volume of liquid displaced by the body

   Ans: d

3. The horizontal force on a curved surface immersed in a liquid is equal to the force on-

   (a) The vertical projected area,

   (b) The horizontal projected area

   (c) Curved surface

   (d) None of the above

   Ans: b

4. The hydrostatic force on a submerged plane surface depends on the __ of the centroid

   (a) Depth,

(b) Width,

(c) Length

(d) None of the above

Ans: a

5. The line of action of the buoyancy force acts through the:

(a) Centre of gravity of any submerged body

(b) Centroid of the volume of any floating body

(c) Centroid of the volume of displaced volume of fluid

(d) None of them

Ans: c

6. A basketball floats in a bathtub of water. The ball has a mass of 0.5 kg and a diameter of 22 cm. What is the buoyant force?

(a) 3.7 N

(b) 2.4 N

(c) 1.8 N

(d) 4.9 N

Ans: d

7. For stable equilibrium of floating bodies of gravity has to be.....

(a) Always below the centre of buoyancy

(b) Always above the centre of buoyancy

(c) Always above the meta centre.

(d) Always below the meta centre.

Ans: d

8. If B is the buoyancy, G is the centre of gravity and M is the metacenter of a floating body, the body will be in stable equilibrium if.

(a) m coincides with G

(b) m is below B

(c) B coincides G

(d) None of them

Ans: a

9. A 2 kg block of wood is floating in water. What is the magnitude of the buoyant force acting on the block?

(a) 19.6 N

(b) 20.5 N

(c) 21.7 N

(d) 18.8 N

Ans: a

10. Which is correct in case of Buoyancy?

(a) The upward force a fluid exerts on an object.

(b) The amount of fluid displaced is equal to the buoyant force pushing up on the object.

(c) If the weight of the fluid displaced is not greater than the object, the object will sink.

(d) All

Ans: d

11. When a ship leaves a river and enters the sea

(a) It will rise a little

(b) It will sink a little

(c) There will be no change in the draft.

(d) It will depend on the type of the ship.

Ans: a

12. When a block of ice floating in water in a container begins to melt the water level in the container

    (a) will rise

    (b) will fall

    (c) will remains constant

    (d) will depend on the shape of the ice block.

    Ans: b

13. An object with specific gravity 4 weighs 100N in air. When it is fully immersed in water its weight will be

    (a) 25 N

    (b) 75 N

    (c) 50 N

    (d) None of the above.

    Ans: b

14. A solid with a specific weight 9020 N/m3 floats in a fluid with a specific weight 10250 N/m^3. The percentage of volume submerged will be

    (a) 90%

    (b) 92%

    (c) 88%

    (d) 78%.

    Ans: c

15. An object weighs 50 N in water. Its volume is 15.3 *l*. Its weight when fully immersed in oil of specific gravity 0.8 will be

    (a) 40 N

    (b) 62.5 N

    (c) 80 N

(d) 65 N.

Ans: c

16. When a small tilt is given to a body floating in stable equilibrium it will ____.

(a) not return to the original position

(b) return to the original position

(c) achieve the new position

(d) None of the above

Ans: b

3.10.2 Theory Questions:

Q. 1 What is the principle of floatation?

Q. 2 Explain stability criterions for a totally submerged body with sketches.

Q. 3 Explain the term 'meta-centre' of a floating body. Derive an expression for the distance between the meta-centre and the centre of buoyancy of a floating body.

Q. 4 Define metacentric height for a floating body and prove that it is given by –

$$GM = \frac{I_{yy}}{V_d} = BG$$

V_d is the volume of liquid displaced by the floating body and BG is the distance between the centre of gravity and centre of buoyancy.

Q. 5 Derive an expression for the total pressure acting on plane surface kept in liquid at angle 'θ' with the free liquid surface. Also determine the location of centre of pressure.

Q. 6 A rectangular block 50 cm long 25 cm wide and 18 cm deep is floating in a liquid the shortest axis of the block is vertical and depth of immersion is 15 cm. Calculate metacentric height and comment on the stability of the block. What will happen if the longest axis is vertical instead of the shortest axis?

Q. 7 Prove that the centre of pressure of a plane surface is always below the centre of gravity. What is the limiting position of the centre of pressure and when does it occur?

Q. 8 Define total pressure and centre of pressure.

Q. 9 Derive the expression for determining the centre of pressure for an inclined triangular plane immersed in water.

3.10.3 Problems:

1. A cylindrical buoy is 2 m in diameter and 2.5 m long and weights 22 kN. The specific weight of sea water is 10.25 kN/m^3. Show that buoy does not float with its axis vertical.

   Ans : BG = 0.9085 m, BM = 0.366 m, GM is negative, The cylinder will not float with its axis vertical.

2. A solid cube of sides 0.5 m each is made of a material of relative density 0.5. The cube floats in a liquid of relative density 0.95 with two of its faces horizontal. Examine its stability.

   Ans: GM = -0.03922 m , Negative sign means meta-centre (m) is below the centre of gravity (G). Thus cube is in unstable equilibrium.

3. A solid cylinder of 200 mm diameter and 800 mm length has its base 20 mm thick and of specific gravity 6. The remaining part of the cylinder is of specific gravity 0.6.

   Ans: GM = -0.03014 m, metacentric height is negative. It means that metacentre is below the C.G. (G). Thus the cylinder is in unstable equilibrium and so it cannot float vertically in water.

4. A frog in a hemispherical pod finds that he just floats without sinking into a sea of blue-green ooze with density 1.35 g/cm^3. If the pod has radius 6 cm and negligible mass, what is the mass of the frog?

   Ans: 610 gm.

5. A rectangular boat made out of concrete with a mass of 3000 kg floats on a freshwater lake (ρ=1000 kg/m3). If the bottom area of the boat is 6 m^2, how much of the boat is submerged?

   Ans: 0.5 m

6. The rock weighs 2.25 newtons when suspended in air. What will be the Buoyant force In water, if weighs 1.8 newtons?

   Ans: 0.45

QR CODES OF SUPPORTING VIDEO LINKS

(1) (2) (3)

REFERENCES AND SUGGESTED READINGS

1. Hydraulics & Fluid Mechanics by Modi and Seth, Standard Book House
2. Theory and Applications of Fluid Mechanics—K. Subramanya- Tata McGraw
3. Fluid Mechanics by R.J. Garde, A.J Mirajgaonkar, SCITECH Publication
4. Fluid Mechanics by Streeter & Wylie, Tata McGraw Hill.
5. https://archive.nptel.ac.in/noc/courses/noc19/SEM2/noc19-ce28.

4 FLUID KINEMATICS

UNIT SPECIFICS

Through this unit we have discussed the following aspects:

- *Fluid flow classification*
- *Streamlines, path lines, streak lines and stream tubes followed by derivation of continuity equation.*
- *Introduction to velocity and acceleration of fluid particles*
- *Rotational and irrotational motion*
- *Concept of velocity potential, stream function and flow net.*

This is followed by large number of solved examples. The students are encouraged to solve the objective questions, long answer questions and numerical problems to judge ones understanding.

RATIONALE

This unit presents kinematical method to determine position, velocity and acceleration of fluid particle at certain time. This is done by using the concept of velocity potential, stream function and flow net. For this one should know the flow classification and flow visualization which is introduced in the initial articles of the unit followed by deriving the continuity equation

PRE-REQUISITES

Mathematics: Derivatives (Class XII)

Physics: Mechanics (Class XII)

UNIT OUTCOMES

List of outcomes of this unit is as follows:

U4-O1: Fluid flow classification

U4-O2: Streamlines, path lines, streak lines, stream tubes

U4-O3: Continuity equation

U4-O4: Rotational and irrotational motion

U4-O5: velocity potential, stream function, flow net

Unit-4 Outcomes	EXPECTED MAPPING WITH COURSE OUTCOMES (1- Weak Correlation; 2- Medium correlation; 3- Strong Correlation)					
	CO-1	CO-2	CO-3	CO-4	CO-5	CO-6
U4-O1	-	-	3	-	-	-
U4-O2	3	-	-	-	-	-
U4-O3	3	-	-	-	-	-
U4-O4	3	-	-	-	-	-
U4-O5	3	-	-	2	-	-

4.1 Introduction:

Kinematics deals with space-time relationship problems of fluids without referring to the forces responsible for this motion. kinematics is solely concerned with the effects of motion on displacement, time, velocity, acceleration or any other quantity derivable from displacement and time (and not the force). Before undertaking the study of forces responsible for motion of fluids, let us first understand the space-time relationship problems.

4.2 Methods Describing Fluid Motion:

There are two important and commonly known methods of describing fluid motion. In 'Lagrangian method' motion of a single fluid particle is studied with respect to the path traced by that fluid particle, its velocity, acceleration etc. over a period of time. In the 'Eulerian method' the motion of fluid particle is observed in a space called 'control volume' to study the velocity, acceleration and position of fluid particles over a period of time. It is usually easier to use the Eulerian approach to describe a flow both in experimental or analytical investigations. A control volume is defined as a volume fixed with respect to co-ordinate system in fluid flow. There are however certain instances in which the Lagrangian method is more convenient. For example, in fluid machinery such as turbines and pumps Lagrangian method is useful as the fluid particles gain or lose energy while moving along their flow paths. Sometimes a coupled Eulerian -Lagrangian technique is adopted to analyze some special problems. In this course we will be using Eulerian technique to study the motion of fluid kinematically. For this it is first necessary to classify the fluid flow with respect to space and time

4.3 Fluid Flow Classification:

The fluid flow may be classified as :

i) Steady and unsteady flow
ii) Uniform and non-uniform flow
iii) Laminar and turbulent flow
iv) Compressible and incompressible flow
v) Rotational and irrotational flow
vi) One,two and three dimensional flow

4.3.1 Steady and Unsteady Flow:

Steady flow is defined as the flow in which the fluid properties like velocity, pressure, density etc. at a point do not change with respect to time. Thus mathematically,

$$\left(\frac{\partial V}{\partial t}\right)_{x_0,y_0,z_0} = 0, \left(\frac{\partial p}{\partial t}\right)_{x_0,y_0,z_0} = 0, \left(\frac{\partial \rho}{\partial t}\right)_{x_0,y_0,z_0} = 0 \quad (4.1)$$

Here (x_0, y_0, z_0) is a fixed point in the fluid.

Unsteady flow is defined as the type of flow in which the velocity, pressure or density etc. at a given point changes with respect to time. Thus,

$$\left(\frac{\partial p}{\partial t}\right)_{x_0,y_0,z_0} \neq 0, \left(\frac{\partial p}{\partial t}\right)_{x_0,y_0,z_0} \neq 0 \text{ etc.} \tag{4.2}$$

However, it must be remembered that as velocity is a vector quantity, the change in its magnitude or direction can make the flow steady.

4.3.2 Uniform and Non-uniform Flow:

Uniform flow is defined as the type of flow in which the fluid characteristics such as velocity, discharge, depth of flow at any given time do not change with respect to space. Mathematically,

$$\left(\frac{\partial V}{\partial S}\right)_{t=\text{ constant}} = 0 \tag{4.3}$$

Where ∂V is change in velocity and ∂S is length of flow in the direction ' S '. Non-uniform flow is the flow in which the fluid characteristics at any given time change with respect to space. Thus,

$$\left(\frac{\partial V}{\partial S}\right)_{t=\text{ constant}} \neq 0 \tag{4.4}$$

4.3.3 Laminar and Turbulent Flow:

Laminar flow is defined as the flow in which the fluid particles move along well-defined paths or in well-defined layers; one layer sliding over the other. This type of flow is also called as stream-line flow or viscous flow. The fluid particles do not move from one layer to other thus moving in orderly manner like marching of solders in army. On the other hand, turbulent flow is characterized by rapid and continuous fluctuations in the velocity components. The fluid particles continuously move from one layer to other and there is no order in their motion like people coming out of cinema theatre.

Whether the flow is laminar or turbulent can be determined by a non-dimensional number $R_e = \frac{VD}{v}$ called as Reynolds Number where V is mean velocity of flow, D is characteristic dimension (diameter in case of flow through circular pipe) and ν is the coefficient of kinematic viscosity of fluid. It is established that if the Reynolds Number is less than 2000 the flow is said to be laminar while the flow is turbulent for Reynolds Number greater than 4000. If the Reynolds Number lies between 2000 to 4000 the flow is in transition state which is very unstable. It may be noted for flow to be classified as Laminar or turbulent the characteristics of flow as explained are important rather than merely the Reynolds number. It has been shown that for Reynolds number greater than 2000 the flow can still exhibit the properties of the laminar flow

4.3.4 Compressible and Incompressible Flow:

Compressible flow is the type of flow in which mass density of the fluid changes because of the application of external pressure. Incompressible flow can be defined as the flow in which mass density of the fluid does not change appreciably because of the application of external pressure. As explained in Unit 1 as in case of the liquids the flow is generally incompressible owing to the incompressible nature of liquids compared to gases.

4.3.5 Rotational and Irrotational Flow:

Rotational flow can be defined as the flow in which fluid particles, rotate about their mass center while flowing along stream lines.

In case of irrotational flows, fluid particles do not rotate about their mass center. A well-known example of irrotational motion is of carriages of the Big (Ferris) wheel in a fair ground. In that big wheel, although each carriage follows a circular path as the wheel revolves, it does not rotate with respect to the earth center, the passenger remains upright and continue to face same direction. Figure 4.1(a) and 4.1(b) represent the irrotational and rotational flow respectively.

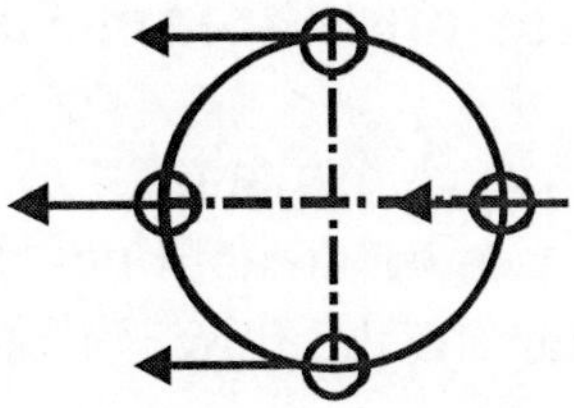

Fig.4.1 (a) Irrotational Flow

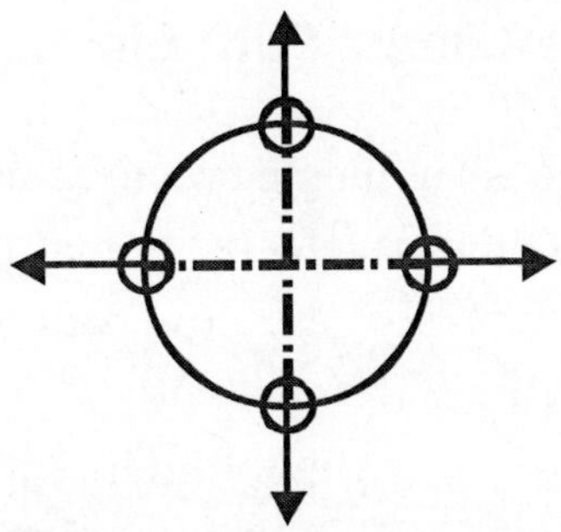

Fig.4.1 (b) Rotational Flow

4.3.6 One, Two and Three-Dimensional Flow:

One dimensional flow can be defined as the flow in which flow parameter such as velocity is a function of time and one-space co-ordinate-only. The variation of velocities in other two mutually perpendicular directions is negligible. Mathematically,

$$u = f(x),\ v = 0 \text{ and } w = 0 \tag{4.5}$$

Here, u, v, w are velocity components in x, y and z directions respectively.

In case of two-dimensional flow velocity is a function of time and two rectangular space co-ordinates while in case of three-dimensional flow velocity is a function of time and three mutually perpendicular directions.

$$u = f(x, y);\ v = f(x, y) \text{ and } w = 0$$

$$u = f(x, y, z);\ v = f(x, y, z),\ w = f(x, y, z) \tag{4.6}$$

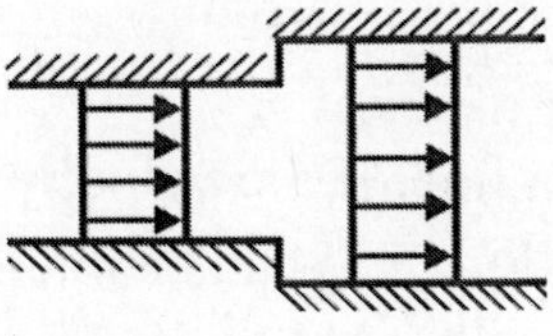

Fig.4.2 (a) 1-D Flow

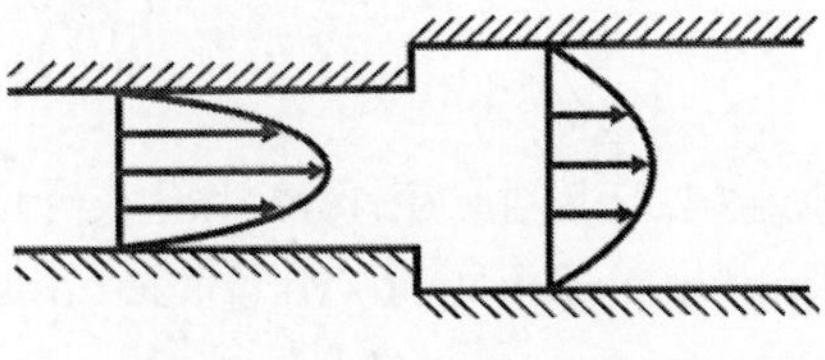

Fig.4.2 (b) 2-D Flow

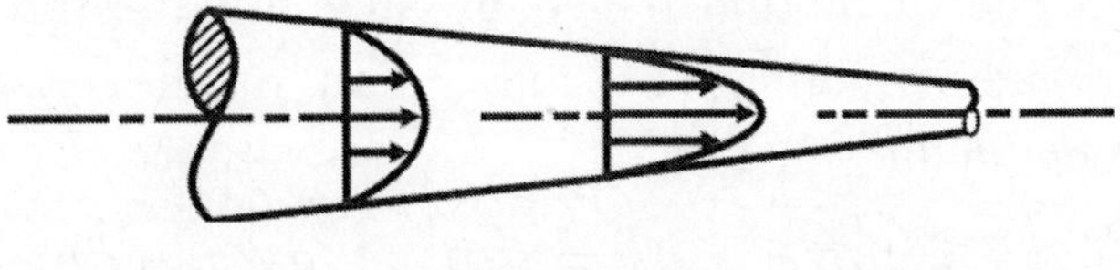

Fig.4.2 (c) 3-D FlowΔ

4.4 Streamlines, Path Lines, Streak Lines and Stream Tubes:

Streamline is an imaginary curve drawn in space such that tangent to it any point gives the direction of velocity at that point. The series of stream lines drawn in this manner will represent the flow pattern at that instant. If the flow is unsteady, the streamline pattern

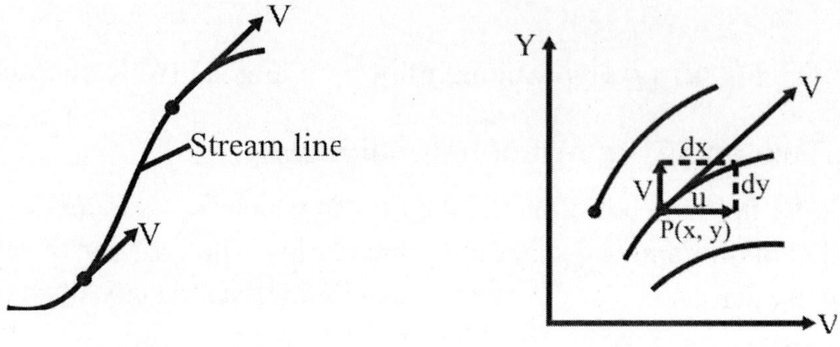

(a) (b)

Fig.4.3 Stream line flow

Consider a two-dimensional flow (figure 4.3b). At any point along a stream line, u and v are the components of tangential velocity V along x and y directions respectively. The slope of the stream line at that point is dy/dx.

mathematically,

$$\left.\begin{aligned}\frac{dy}{dx} &= \frac{v}{u} \\ udy - vdx &= 0\end{aligned}\right\} \quad (4.7)$$

Equation (4.7) is the differential equation of a stream line. If the velocity components are known as functions of space co-ordinates, equation (4.1) can be integrated to give the velocity of the fluid at that instant kinematically without referring to the force responsible for the motion.

Path line is the line of motion traced by an individual fluid particle in a finite time. The path line is defined by integration of the relation between velocity and displacement in the equation,

$$dx = udt, \quad dy = vdt, \quad dz = wdt \quad (4.8)$$

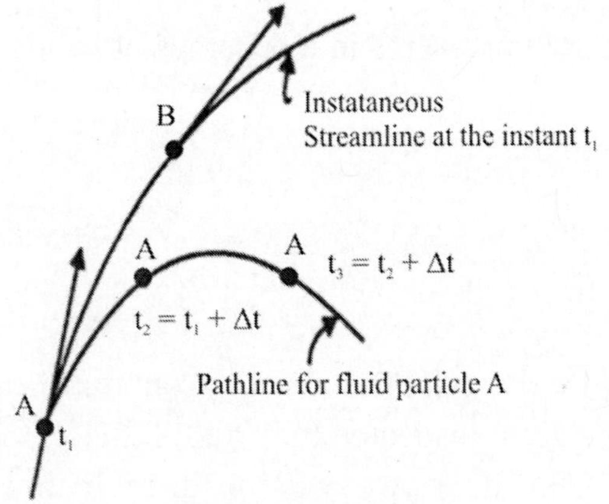

Fig.4.4 Streamline & pathline for unsteady non uniform flow

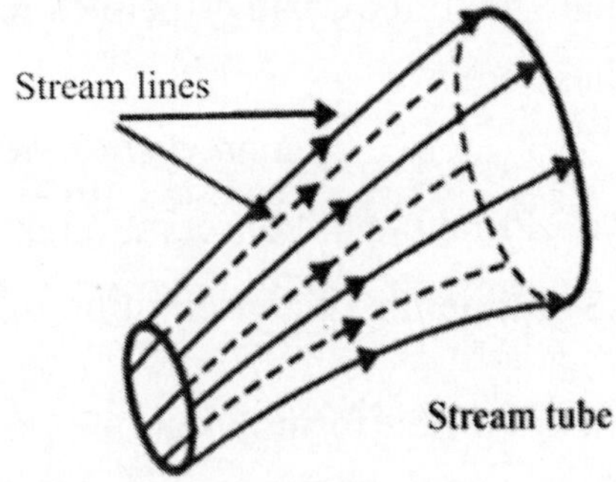

Fig. 4.5 Stream tube

For steady flow, the particle velocity at any point in the flow field will not depend on time and all particles that pass through this point will follow the same trajectories. Thus path lines are identical to stream lines for steady flow (refer figure 4.5)

A streak line is the locus of particles which have passed through a fixed point. In other words, a streak line is an instantaneous locus of all fluid particles that have passed through a given point, i.e .streak lines give the spread of the fluid particles in the space. In experimental work, streak lines are obtained by injecting dye or smoke into the flowing fluid. The resulting-colored lines are the streak lines.

For steady flow path lines, stream lines and streak lines coincide with each other.

Stream tube can be defined as a group of stream lines. The bound surface of the stream tube is made up of several stream lines. The property of stream line makes the concept of stream tube very useful. Since at any point along the stream-line the velocity is tangential to the stream line; the component of velocity at right angles to the stream line is always zero. Therefore, there is no flow across the stream line. As the surface of stream tube is made up of numerous stream lines, one can conclude that there is no flow across the surface of a stream tube.

It may be understood that the streamline, path line, streak line, stream tube are imaginary and do not exist physically in any flow.

4.5 Principle of Conservation of Mass:

There are three basic principles for analyzing the problems in mechanics of solids. These are:

i) Principle of conservation of mass

ii) Principle of conservation of energy

iii) Principle of conservation of momentum.

The modified forms of above principles in fluid mechanics are the Continuity equation, Energy equation and Momentum equation respectively. Out of these three are continuity equation is derived here. The other two will be dealt with in Fluid Dynamics.

4.5.1 Continuity Equation in Cartesian Co-ordinates:

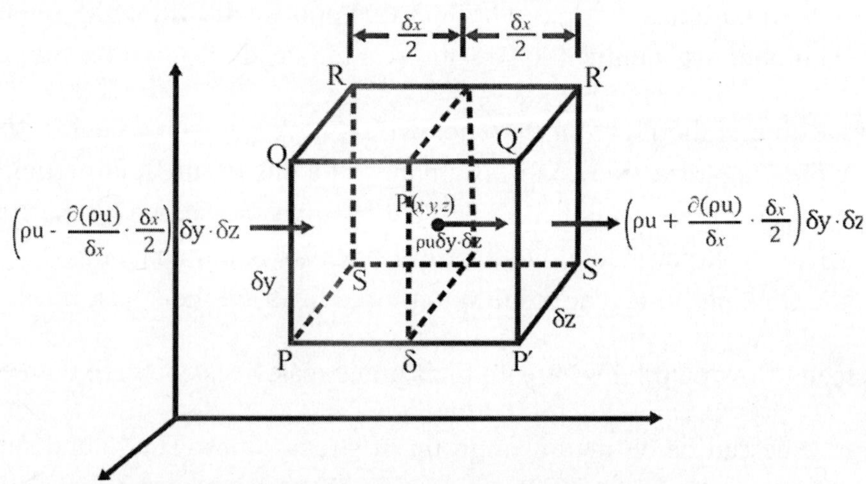

Fig.4.6 Continuity Equation -Elementary parallelopiped

Continuity equation expresses the fact that mass can neither be created nor be destroyed. In fluid mechanics mass is always expressed as mass per time for obvious reasons. The mass is the product of volume and density. Volume per time is called as discharge (Q=V/A, defined later) making the product of density and discharge as mass per time. (mass= ρ Q). The discharge in turn will be calculated as product of velocity and area

(Q = v A explained later). The conservation of mass thus takes the form of continuity of discharge in fluid mechanics

Consider an elementary parallelopiped with sides ∂x, ∂y and ∂z respectively. Refer figure 4.6.

Let P(x, y, z) be the centre of parallelopiped and let u, v and w represent velocities in x, y, z directions respectively at point P. Let ρ represents the mass density of the fluid. Then the mass of fluid passing through area (δyδz) through point P is ($\rho u\delta y\delta z$). Let us assume that (ρu) varies along x axis as u is velocity component along x axis.

The mass flowing through face PQRS per unit time will be

$$= \left\{\rho u - \frac{\partial}{\partial x}(\rho u)\frac{\delta x}{2}\right\}\delta y\delta z \qquad (4.9)$$

Here it is to be understood that the mass (ρuδyδz) is varying (decreasing) along x axis for a distance equal to half the length

$$\left(\frac{\partial}{\partial x}(\rho u\delta y\delta z)\cdot\frac{\delta x}{2}\right)$$

on the left hand side of point P while on the right hand side it increases. Therefore, the mass flowing out through face P′Q′R′S′ per unit time will be

$$= \rho u\delta y\delta z + \frac{\partial}{\partial x}(\rho u\delta y\delta z)\cdot\frac{\delta x}{2}$$

$$= \left\{\rho u + \frac{\partial}{\partial x}(\rho u)\frac{\delta x}{2}\right\}\delta y\delta z \qquad (4.10)$$

The net inflow of mass through faces PQRS and P′Q′R′S′ is obtained by subtracting Equation (4.3) from Equation (4.4) i.e.

$$= \left\{(\rho u) - \frac{\partial}{\partial x}(\rho u)\frac{\delta x}{2} - (\rho u) - \frac{\partial}{\partial x}(\rho u)\frac{\delta x}{2}\right\}\delta y\delta z$$

$$= -\frac{\partial}{\partial x}(\rho u)\,\delta x\delta y\delta z \qquad (4.11)$$

Similarly, the net mass inflow through other two pairs of faces can be written as:

$$= -\frac{\partial}{\partial y}(\rho v)\delta x \delta y \delta z$$

$$= -\frac{\partial}{\partial z}(\rho w)\, \delta x \delta y \delta z \qquad (4.12)$$

Therefore, net mass inflow in the parallelopiped is summation of Equations (4.11) and (4.12).

$$\text{Mass of fluid in parallelopiped} = \rho\ \delta x\ \delta y\ \delta z \qquad (4.13)$$

$$\text{Rate of change of mass in the parallelepiped} = \frac{\partial}{\partial t}(\rho\ \delta x\ \delta y\ \delta z) \qquad (4.14)$$

Obviously, the net mass inflow into the parallelopiped through all the faces must be equal to the rate of change of mass of the parallelopiped. Thus,

$$\frac{\partial(\rho \delta x \delta y \delta z)}{\partial t} = \left\{-\frac{\partial(\rho u)}{\partial x} - \frac{\partial(\rho v)}{\partial y} - \frac{\partial(\rho w)}{\partial z}\right\} \delta x \delta y \delta z \qquad (4.15)$$

Dividing both sides of Equation (4.15) by volume of parallelepiped $\delta x \delta y \delta z$

$$\frac{\partial \rho}{\partial t} + \frac{\partial \rho u}{\partial x} + \frac{\partial \rho v}{\partial y} + \frac{\partial \rho w}{\partial z} = 0 \qquad (4.16)$$

This is the continuity equation for three-dimensional flow steady, unsteady, uniform, non-uniform as well as compressible and incompressible flow in cartesian coordinates

For steady flow $$\frac{\partial \rho}{\partial \mathrm{t}} = 0 \qquad (4.17)$$

Therefore, continuity equation for steady flow is,

$$\frac{\partial(\rho u)}{\partial x} + \frac{\partial(\rho v)}{\partial y} + \frac{\partial(\rho w)}{\partial z} = 0 \qquad (4.18)$$

For incompressible fluids ρ is constant. Therefore, continuity equation for steady-incompressible flow will be

$$\frac{\partial u}{\partial x}+\frac{\partial v}{\partial y}+\frac{\partial w}{\partial z}=0 \tag{4.19}$$

For two dimensional flow,

$$\frac{\partial u}{\partial x}+\frac{\partial v}{\partial y}=0 \tag{4.20}$$

4.5.2 Continuity Equation Based on Stream Tube Concept:

Consider an elementary tubular shaped control volume of fluid along a stream tube. Refer Fig. 4.7.

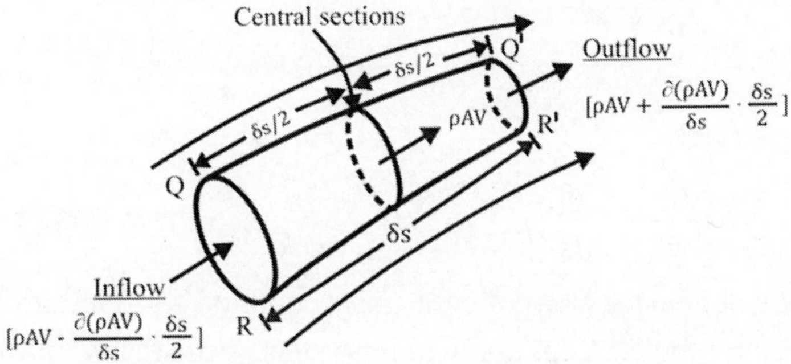

Fig.4.7 Flow through a stream tube

The flow takes place through the ends of the element and not through the surface of the control volume (refer stream-tube). Let A, V and ρ be the area of cross section, velocity and mass density of fluid at the central section of the element respectively, and A, V and ρ are function of s only.

Rate of inflow of mass in the element through RQ

$$=\rho \text{AV}-\frac{\partial(\rho \text{AV})}{\partial s}\cdot\frac{\delta s}{2} \tag{4.21}$$

Rate of fluid flowing out of the element through $R'Q'$

$$=\rho \text{AV}+\frac{\partial(\rho \text{AV})}{\partial s}\cdot\frac{\delta s}{2} \tag{4.22}$$

Therefore,

net rate of increase of mass in the control volume

$$= -\frac{\partial}{\partial s}(\rho \mathrm{AV}) \,.\, \delta s \tag{4.23}$$

Net mass of fluid which can be accommodated in the control volume

$$= \rho \mathrm{A} \delta \mathrm{s} \tag{4.24}$$

Therefore, rate of change of mass in the control volume per unit time

$$= \frac{\partial}{\partial t}(\rho . \mathrm{A} \delta \mathrm{s}) \tag{4.25}$$

By conservation of mass principal, equating equations (4.24) and (4.25),

$$\frac{\partial}{\partial t}(\rho \mathrm{A} \delta \mathrm{s}) = -\frac{\partial}{\partial s}(\rho \mathrm{A} V) \cdot \delta \mathrm{s} \tag{4.26}$$

simplifying we get,

$$\frac{\partial \rho A}{\partial t} + \frac{\partial}{\partial s}(\rho \mathrm{A} V) = 0 \tag{4.27}$$

This is the most general form of continuity equation for one dimensional flow and is based on stream tube concept applicable to steady, unsteady, uniform, non-uniform as well as compressible and incompressible flows.

For steady flow, variation of parameters of flow with respect to time are zero.

$$\therefore \frac{\partial \rho A}{\partial t} = 0 \tag{4.28}$$

∴ Equation (4.21) can be written as,

$$\frac{\partial}{\partial s}(\rho \mathrm{A} V) = 0 \text{ or } \rho \mathrm{A} V = \text{Constant} \tag{4.29}$$

or for different cross-sections of stream tube

$$\rho_1 A_1 V_1 = \rho_2 A_2 V_2 = \rho_3 A_3 V_3 = \rho_n A_n V_n \tag{4.30}$$

This is the continuity equation for steady flow in one dimension applicable to compressible as well as incompressible flow. For incompressible fluid ρ is constant.

Therefore, continuity equation for steady, one-dimensional flow of an incompressible fluid can be written as,

$$\frac{\partial}{\partial s}(\text{AV}) = 0$$

$$\therefore \text{AV} = \text{Constant}$$

$$\text{or } A_1V_1 = A_2V_2 = A_3V_3 \tag{4.31}$$

The product AV represents the volume of fluid passing the control volume per second and is called as discharge and is denoted by Q.

Q = AV is discharge having dimensions $[L^3T^{-J}]$.

4.6 Velocity and Acceleration of Fluid Particles:

Velocity is always tangential to the path of the fluid particle and is a function of space and time. V = $f(x, y, z, t)$. It can be resolved in three components u, v, w along three coordinates axes X,Y,Z axes respectively. It is defined as the distance traveled per time V = ds/dt where ds is distance travelled by fluid in time dt.

The general expression for acceleration can be obtained by taking total differentials of expressions for u, v and w. Since $u = u(x, y, z, t)$, one can write the expression for total change in u as,

$$= \frac{\partial u}{\partial x}dx + \frac{\partial u}{\partial y}dy + \frac{\partial u}{\partial z}dz + \frac{\partial u}{\partial t}dt \tag{4.32}$$

However, $dx = udt,\ dy = vdt,\ dz = wdt$

$$\therefore du = u\frac{\partial u}{\partial x}dt + v\frac{\partial u}{\partial y}dt + w\frac{\partial u}{\partial z}dt + \frac{\partial u}{\partial t}dt \tag{4.33}$$

$$\therefore a_x = \frac{\partial u}{\partial t} = u\frac{\partial u}{\partial x}dt + v\frac{\partial u}{\partial y}dt + w\frac{\partial u}{\partial z}dt + \frac{\partial u}{\partial t}dt \tag{4.34}$$

Here, a_x is the Total acceleration in x direction. Similarly the expression obtained for a_y and a_z.

$$\therefore a_y = \frac{dv}{dt} = u\frac{\partial v}{\partial x}dt + v\frac{dv}{\partial y}dt + w\frac{dv}{\partial z}dt + \frac{dv}{\partial t}dt \tag{4.34}$$

$$\therefore a_y = \frac{dv}{dt} = u\frac{\partial v}{\partial x}dt + v\frac{dv}{\partial y}dt + w\frac{dv}{\partial z}dt + \frac{dv}{\partial t}dt \tag{4.35}$$

The first three terms in Equations (4.33), (4.34) and (4.35) are called as convective acceleration while last term represents local acceleration. The above equations can be reduced to 2 - D or 1 - D form as the flow case may be.

4.7 Rotational and Irrotational Motion:

During course of movement in the direction of flow, a fluid particle may undergo any one or combination of following types of motion

(i) Linear translation or pure translation

(ii) Linear deformation

(iii) Angular deformation

(iv) Rotation

Linear translation is defined as the movement of a fluid element in such a way that it moves bodily from one position to another position and the two axes ab and cd represented in new positions by $a'b'$ and $c'd'$ are parallel as shown in figure 4.8 (a).

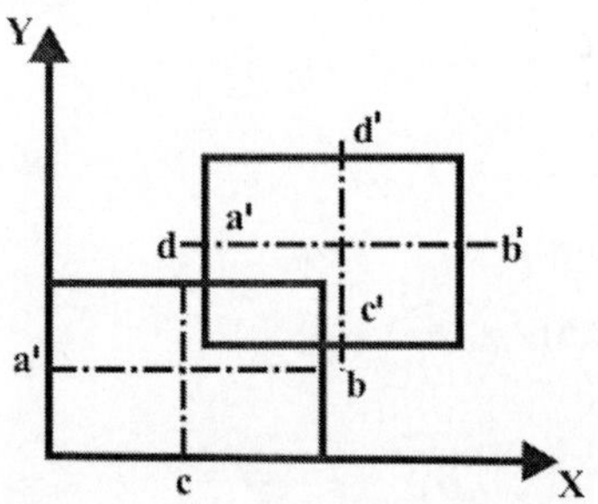

(a)

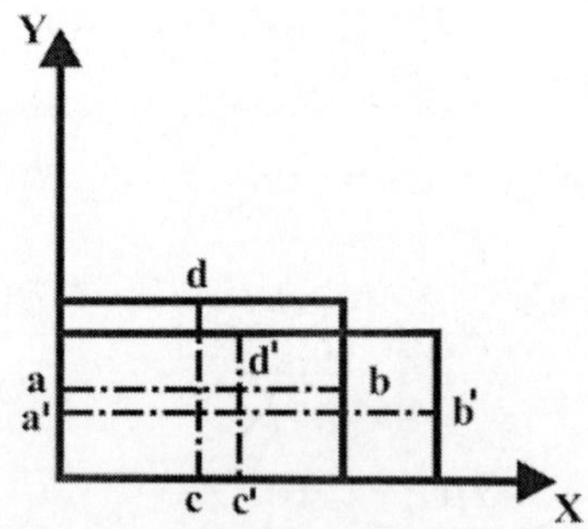

(b)

Fig.4.8 (a) and (b) Rotational and irrotational motion

Linear deformation is defined as the deformation of a fluid element in linear direction when the element moves. The axes of the element in the deformed position and undeformed position are parallel but their lengths change as shown in figure 4.8 (b).

Angular deformation or shear deformation is defined as the average change in the angle contained by two adjacent sides. Let $\Delta\theta_1$ and $\Delta\theta_2$ is the change in angle between two adjacent sides of a fluid element as shown in figure 4.9(a), then, angular deformation or

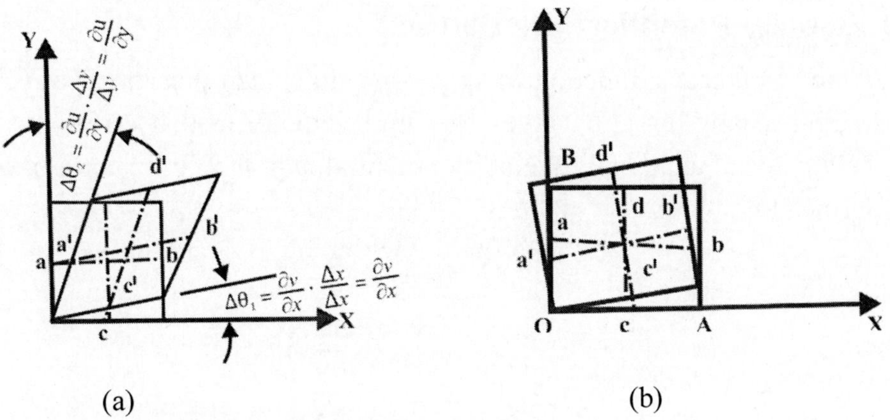

Fig.4.9 (a and b) Rotational motion

$$\text{shear strain rate} = \frac{1}{2}[\Delta\theta_1 + \Delta\theta_2] \tag{4.36}$$

$$\Delta\theta_1 = \frac{\partial v}{\partial x} \cdot \frac{\Delta x}{\Delta x} = \frac{\partial v}{\partial x} \text{ and } \Delta\theta_2 = \frac{\partial u}{\partial y} \cdot \frac{\Delta y}{\Delta y} = \frac{\partial u}{\partial y} \tag{4.37}$$

$$\text{Angular deformation} = \frac{1}{2}[\Delta\theta_1 + \Delta\theta_2]$$

$$\text{or shear strain rate} = \frac{1}{2}\left[\frac{\partial v}{\partial x} + \frac{\partial u}{\partial y}\right] \tag{4.38}$$

Rotation is defined as the movement of a fluid element in such a way that both of its horizontal as well as vertical axes rotate in the same direction as shown in figure 4.9 (b). It is equal to $\frac{1}{2}\left(\frac{\partial v}{\partial x} - \frac{\partial u}{\partial y}\right)$ for a two-dimensional element in x-y plane. The rotational components are

$$\left.\begin{aligned} \omega_x &= \frac{1}{2}\left(\frac{\partial v}{\partial x} - \frac{\partial u}{\partial y}\right) \\ \omega_y &= \frac{1}{2}\left(\frac{\partial \omega}{\partial y} - \frac{\partial v}{\partial z}\right) \\ \omega_z &= \frac{1}{2}\left(\frac{\partial u}{\partial z} - \frac{\partial \omega}{\partial x}\right) \end{aligned}\right\} \tag{4.39}$$

Vorticity is defined as the value twice of the rotation and hence it is given as 2ω.

4.8 Velocity Potential Function and Stream Function:

4.8.1 Velocity Potential Function ϕ:

It is defined as a scalar function of space and time such that its negative derivative with respect to any direction gives the fluid velocity in that direction. It is defined by $\phi(Phi)$. Mathematically, velocity potential $\phi = f(x, y, z)$ for a steady, 3-D flow is defined as,

$$\left.\begin{aligned} u &= -\frac{\partial\phi}{\partial x} \\ v &= -\frac{\partial\phi}{\partial y} \\ w &= -\frac{\partial\phi}{\partial z} \end{aligned}\right\} \tag{4.40}$$

Hereu, v, w are components of velocity in x, y and z directions respectively. The negative sign signifies that the flow takes place in the direction of decreasing velocity potential.

For a 2-D continuity equation,

$$\frac{\partial \mathrm{u}}{\partial x} + \frac{dv}{\partial \mathrm{y}} = 0$$

Substituting the values of u and v from Equation (4.40)

$$\frac{\partial}{\partial x}\left(-\frac{\partial\phi}{\partial x}\right) + \frac{\partial}{\partial y}\left(-\frac{\partial\phi}{\partial y}\right) = 0$$

$$\frac{\partial^2\phi}{\partial x^2} + \frac{\partial^2\phi}{\partial y^2} = 0 \tag{4.41}$$

Equation (4.41) is called as Laplace equation thus $\Delta^2 \phi = 0$

Thus, any value of ϕ that satisfies Laplace's equation will correspond to some case of fluid flow. Similarly substituting of u and v in expressions of rotational component.

$$\omega_x = \frac{1}{2}\left(\frac{\partial v}{\partial x} - \frac{\partial u}{\partial y}\right) \tag{4.42}$$

$$\omega_y = \frac{1}{2}\left(\frac{\partial u}{\partial z} - \frac{\partial \omega}{\partial y}\right)$$

We get,

$$\omega_x = \frac{1}{2}\left[\frac{\partial}{\partial x}\left(-\frac{\partial \phi}{\partial x}\right) - \frac{\partial}{\partial y}\left(-\frac{\partial \phi}{\partial x}\right)\right] = \frac{1}{2}\left[-\frac{\partial^2 \phi}{\partial x\,\partial y} + \frac{\partial^2 \phi}{\partial y\,\partial x}\right] \tag{4.43}$$

$$\omega_x = \frac{1}{2}\left[\frac{\partial}{\partial z}\left(-\frac{\partial \phi}{\partial x}\right) - \frac{\partial}{\partial x}\left(-\frac{\partial \phi}{\partial z}\right)\right] = \frac{1}{2}\left[-\frac{\partial^2 \phi}{\partial z\,\partial x} + \frac{\partial^2 \phi}{\partial x\,\partial z}\right] \tag{4.43}$$

If ϕ is a continuous function then,

$$\frac{\partial^2 \phi}{\partial x\,\partial y} = \frac{\partial^2 \phi}{\partial y\,\partial x}; \frac{\partial^2 \phi}{\partial z\,\partial x} = \frac{\partial^2 \phi}{\partial x\,\partial z} \tag{4.44}$$

$$\therefore \quad \omega_x = \omega_y = 0$$

Thus, if the rotational components are zero then that flow is called as irrotational flow. Hence the properties of a potential functions are as follows:

(i) If velocity potential (ϕ) exist, the flow is irrotational. (ii) If velocity potential (ϕ) satisfies the Laplace's equation, it represents the possible case of steady incompressible irrotational flow.

4.8.2 Stream Function ψ:

It is defined as the scalar function of space and time, such that its partial derivative with respect to any direction gives the velocity component at right angles to that direction (in anticlockwise sense). It is denoted by ψ (Psi) and is defined only for two-dimensional flow. Mathematically, for steady flow $\psi = f(\psi x, y)$ is defined as,

$$\left.\begin{aligned}\frac{\partial \psi}{\partial x} &= v \\ \frac{\partial \psi}{\partial y} &= -u\end{aligned}\right\} \tag{4.45}$$

The continuity equation for 2-D flow is

$$\frac{\partial \mathrm{u}}{\partial x} + \frac{\partial \mathrm{v}}{\partial \mathrm{y}} = 0 \tag{4.46}$$

Substituting the values of u and v from the Equation (4.35) we get,

$$\frac{\partial}{\partial x}\left(-\frac{\partial \psi}{\partial y}\right) + \frac{\partial}{\partial y}\left(\frac{\partial \psi}{\partial x}\right) = 0$$

$$\text{Or} \quad -\frac{\partial^2 \psi}{\partial x\, \partial y} + \frac{\partial^2 \psi}{\partial x\, \partial y} = 0 \tag{4.47}$$

Hence existence of ψ also means a possible case of fluid flow. The flow may be rotational or irrotational. The rotational component ω_Z is given by,

$$\omega_Z = \frac{1}{2}\left(\frac{\partial \mathrm{v}}{\partial x} - \frac{\partial \mathrm{u}}{\partial \mathrm{y}}\right)$$

Substituting the values of u and v from the Equation (4.33) in the above equation

$$\omega_Z = \frac{1}{2}\left[\frac{\partial}{\partial x}\left(\frac{\partial \psi}{\partial x}\right) - \frac{\partial}{\partial \mathrm{y}}\left(-\frac{\partial \psi}{\partial \mathrm{y}}\right)\right]$$

$$= \frac{1}{2}\left[\frac{\partial^2 \psi}{\partial x^2} + \frac{\partial^2 \psi}{\partial \mathrm{y}^2}\right] \tag{4.48}$$

For an irrotational flow $\omega_z = 0$.

thus $\Delta^2\, \psi = 0$, which is Laplace's equation for ψ.

The properties of stream function ψ are as follows:

(i) If stream function (ψ) exists, it is a possible case of fluid flow which may be rotational or irrotational.

(ii) If stream function (ψ) satisfies the Laplace equation, it is a possible case of irrotational flow.

4.8.3 Equipotential Lines:

A line along which the velocity potential ϕ is constant, is called as equipotential line. For equipotential line ϕ=constant.

$$\therefore d\phi = 0 \tag{4.49}$$

But $\phi = f(x, y)$ for steady flow

$$\therefore d\phi = \frac{\partial \phi}{\partial x} dx + \frac{\partial \phi}{\partial y} dy$$

$$= -udx - vdy \qquad \left\{\because \frac{\partial \phi}{\partial x} = -u \, \frac{\partial \phi}{\partial y} = -v\right\}$$

$$d\phi = -(udx + vdy) \tag{4.50}$$

For equipotential line $d\phi = 0$

$$\therefore -(udx + vdy) = 0 \;\text{ or }\; udx + vdy = 0$$

$$\frac{dy}{dx} = -\frac{u}{v} \tag{4.51}$$

Equation 4.51 is the equation of equipotential line wherein dy/dx is the slope of equipotential line.

4.8.4 Line of Constant Stream Function:

The line along which the value of stream function is constant is called as streamline. i.e. ψ = constant

$$\therefore d\psi = 0$$

$$d\psi = \frac{\partial \psi}{\partial x} dx + \frac{\partial \psi}{\partial y} dy \;\text{ as }\; \psi = f(x,y) \tag{4.52}$$

$$= v\mathrm{d}x - u\mathrm{d}y \qquad \because \frac{\partial\psi}{\partial x} = v; \ \frac{\partial\psi}{\partial y} = -u$$

For a line of constant stream function

$$\mathrm{d}\psi = 0 \ \text{ or } \ v\mathrm{d}x - u\mathrm{d}y = 0$$

$$\therefore \qquad \frac{\partial y}{\partial x} = \frac{v}{u} \tag{4.53}$$

Equation 4.53 is equation of a streamline wherein $\frac{\partial y}{\partial x}$ represents slope of the line for which stream function is constant.

From Equations (4.51) and (4.53) it is clear that the product of the slope of the equipotential line and the slope of the streamline at the point of intersection is equal to -1. Thus the equipotential lines are orthogonal to streamlines at all points of intersection.

Considering the definitions and velocity potential and stream function and equations 4.39 and 4.42

$$\left.\begin{aligned} u &= -\frac{\partial\phi}{\partial x} \\ v &= -\frac{\partial\phi}{\partial y} \\ w &= -\frac{\partial\phi}{\partial z} \end{aligned}\right\} \text{ and } \begin{aligned} \frac{\partial\psi}{\partial x} &= v \\ \frac{\partial\psi}{\partial y} &= -u \end{aligned}$$

It can be proved that

$$\frac{\partial\phi}{\partial x} = \frac{\partial\psi}{\partial y}$$

$$-\frac{\partial\phi}{\partial y} = \frac{\partial\psi}{\partial x} \tag{4.54}$$

Equations 4.54 are known as Cauchy-Riemann equations

4.9 Flow net:

A grid obtained by drawing a series of streamlines and equipotential lines is known as flow net. Obtaining a flow pattern or flow net for steady two-dimensional irrotational flow involves solution of Laplace equation with given boundary conditions.

For any potential flow field, a flow net can be drawn that consists of family of streamlines and equipotential lines. The flow net is useful in visualizing flow patterns and can be used to obtain graphical solution by sketching the streamlines and equipotential lines and adjusting the lines until the lines are approximately orthogonal at all points where they intersect. Figure 4.10 shows a network of mutually perpendicular streamlines and equipotential lines.

The streamlines which show the direction of flow at any point are so spaced that there is an equal rate of flow dq discharging through each tube. The discharge dq is equal to the change in ψ from one streamline to the next. The equipotential lines are then drawn everywhere normal to the streamlines.

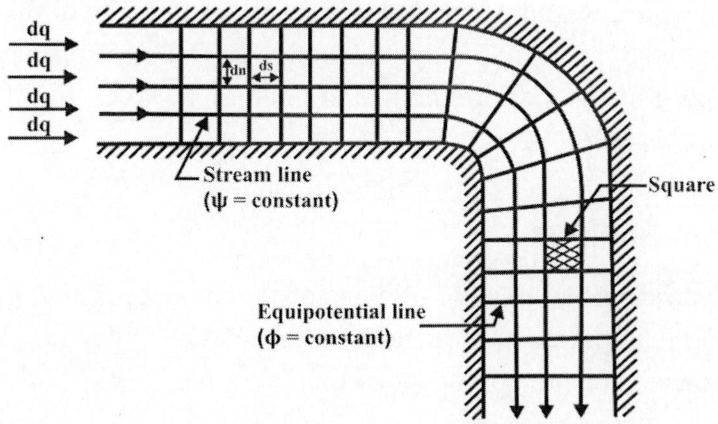

Fig. 4.10 Flow net within a closed conduct

The spacing of the equipotential line is selected in such a way that the change in velocity potential from one equipotential line to the next is constant. Furthermore, the changes in both sets of lines are made equal. In other words, $d\psi = d\phi$. Therefore, they form approximate squares.

These are some specific characteristics of flow net:

(i) Flow nets can be drawn for irrotational flows.

(ii) The streamlines intersect equipotential lines including fixed boundaries normally.

(iii) There is only one possible pattern of flow net for a given set of boundary conditions if correctly prepared will represent this pattern.

4.9.1 Methods of Drawing Flow Net:

There are following methods of drawing the methods:

(i) Graphical method

(ii) Numerical method

(iii) Electrical analogy method.

Graphical method is comparatively simple and may be used to determine the flow pattern for any given set of boundary conditions.

Numerical methods are based upon the calculus of finite differences used to determine patterns of flow in cases in which, because of the complexity of their boundary forms cannot be solved by ordinary analytical methods. The electrical analogy method is a general-purpose method where the use of analogy between flow of "electricity and flow of fluids" is used.

4.9.1.1 Electrical Analogy:

The flow of an electric current in a two-dimensional conductor is analogous to irrotational flow. The electric potential is analogous to velocity potential. The homogeneous conductor is analogous to homogenous fluid.

Fig.4.11 shows a simple arrangement for an electrical analogy study.

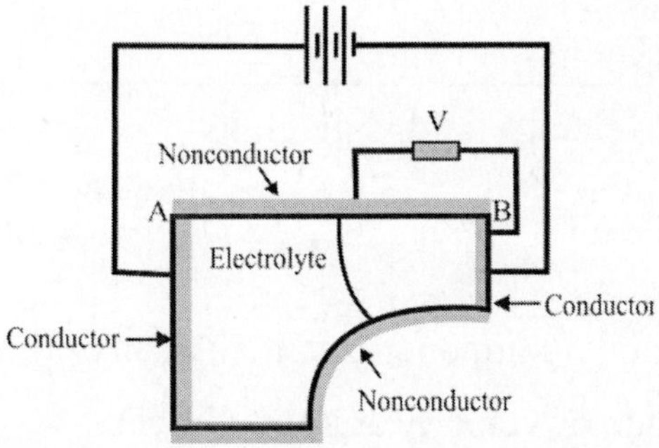

Fig.4.11 A simple arrangement for an electrical analogy study

The fixed boundaries of a hydraulic model are replaced by conductors and non-conductors. Conductors are fixed in the direction of flow. The container thus formed is filled with an electrolyte such as salt-water. Electric potential difference is developed between the two conductors. With the help of probe of null indicator of voltmeter, the points having equal potential are found out. The points having equal potential if joined with smooth curves will indicate equipotential lines. Once equipotential lines are fixed stream lines can be drawn graphically. Set of equipotential lines and stream lines give the flow net.

4.9.2 Uses of Flow Net:

i. Flow net may be used for all irrotational flows with geometrically similar conditions.

ii. Once the flow is drawn, the spacing between the adjacent streamlines is determined and the application of the continuity equation gives the velocity of flow at any point, if the velocity of flow at any reference point is known.

iii. The flow net analysis assists in the determination of the efficient boundary shapes for which the flow does not separate from the boundary surface.

iv. The flow cases where viscosity effects are comparatively unimportant, the flow net analysis can be effectively used.

4.10 Solved Examples:

Ex. 4.1: In a fluid, the velocity field is given by

$$V = (3x + 2y)i + (2z + 3x^2)j + 2t - 3z)k$$

Determine:

(i) The velocity components u, v, w at any point in the flow field.

(ii) Magnitude of velocity at point (1,1,1)

(iii) Magnitude of velocity at $t = 2$ and at point (2,2,1).

Also classify the velocity field as steady or unsteady, uniform or non-uniform and one, two or three dimensional.

Solution :

$$u = 3x + 2yv = 2z + 3x^2, \quad w = 2t - 3z, \quad \text{at (1,1,1)}$$

$$u = 3 + 2 = 5$$

$$v = 2 + 3 = 5$$

$$w = 2t - 3$$

$$\therefore |V| = \sqrt{u^2 + v^2 + w^2}$$

$$= \sqrt{25 + 25 + (2t - 3)^2}$$

$$= \sqrt{50 + 4t^2 - 12t + 9}$$

$$|V|_{(1,1,1)} = \sqrt{4t^2 - 12t + 59}$$

$$\text{at } t = 2 \text{ and } x = 2, y = 2, z = 1$$

$$u = 6 + 4 = 10$$

$$v = 2 + 12 = 14$$

$$w = 1$$

$$\therefore |V| = \sqrt{u^2 + v^2 + w^2}$$

$$= \sqrt{100 + 156 + 1} = \sqrt{297}$$

$$= 18.89 \text{ units}$$

As V depends on t the flow is unsteady.

The velocity changes in x direction at given t. Therefore, flow is non-uniform and V depends on x, y, z. Therefore, it is three dimensional.

Ex. 4.2: In a three-dimensional fluid flow two velocity components u' and v are $u' = 2x^2$ and $v = 2xyz$. Find the third component w such that the continuity equation is satisfied.

Solution:

continuity equation for steady flow is,

$$\frac{\partial u}{\partial x} + \frac{\partial v}{\partial y} + \frac{\partial w}{\partial z} = 0$$

$$4x + 2xz + \frac{\partial w}{\partial z} = 0$$

$$\frac{\partial w}{\partial z} = -\ (4x + 2xz)$$

Integrate Equation (1) w.r.t. to z,

$$w = -\ [4xz + xz^2] + C \qquad \text{where, } C = f(x,y)$$

Ex. 4.3: In a two-dimensional fluid motion stream function is given by $\psi = x^2 - y^2$.

1 Determine the velocity and its direction at (21, 2).
2 Sketch the stream lines and show the direction of flow.

Solution:

$$\frac{\partial \psi}{\partial x} = v, \qquad \frac{\partial \psi}{\partial y} = -u$$

$$2x = v, \qquad -2y = -u$$

$v = 42$ units $\qquad u = 4$ units

$$|V| = \sqrt{v^2 + u^2} = \sqrt{1780} = 42.19 \text{ units}$$

$$\theta = \tan^{-1}\frac{v}{u} = \tan^{-1}\frac{42}{4}$$

Equation $\psi = x^2 - y^2$ represents a hyperbola.

The table gives values of x and y for some fixed values of ψ.

ψ	y	$x = \pm\sqrt{y^2 + \psi}$
1	0	± 1
	1	$\pm\sqrt{2}$
2	0	$\pm\sqrt{5}$
	1	$\pm\sqrt{2}$
	2	$\pm\sqrt{6}$

Thus stream lines can be represented as,

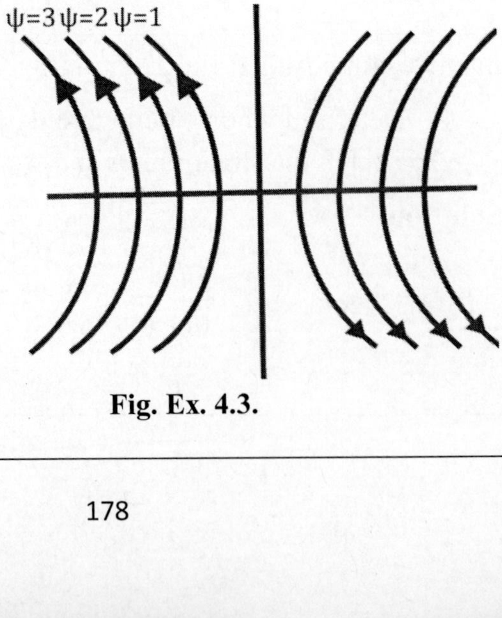

Fig. Ex. 4.3.

Ex. 4.4:	A flow has a potential function ϕ given by $\phi = (x^3 - 3xy^2)$. Derive the corresponding stream function ψ and evaluate magnitude and direction of velocity at any arbitrary point x, y. **Solution. :** $$\phi = (x^3 - 3xy^2)$$ $$\frac{\partial \phi}{\partial x} = 3x^2 - 3y^2 = 3(x^2 - y^2) = -u$$ $$\frac{\partial \phi}{\partial y} = 6xy = -v$$ but $\frac{\partial \psi}{\partial x} = v$ from definition of ψ $$\frac{\partial \psi}{\partial x} = 6xy$$ Integrating, with respect to x $$\psi = 6y\frac{x^2}{2} + f(y)$$ $$= 3x^2 y + f'(y)$$ $$\therefore \frac{\partial \psi}{\partial y} = 3x^2 + f'(y)$$ but $\frac{\partial \psi}{\partial y}$ = $-u$ by definition which is equal to $\frac{\partial \phi}{\partial x}$ $$\therefore \frac{\partial \psi}{\partial y} = -u = 3(x^2 - y^2)$$ $$= 3(x^2 - y^2)$$ Comparing Equation (1) and Equation (2) $$f'(y) = -3y^2$$

Integrating

$$f(y) = -y^3 + c$$

$$\therefore \psi = 3x^2y - y^3 + c$$

$$\Rightarrow \psi = y(3x^2 - y^2) + c$$

as passing through origin, $\therefore$ c = 0.

$$\psi = y(3x^2 - y^2)$$

$$u = 3(y^2 - x^2)$$

$$v = 6xy$$

$$= \sqrt{9(y^2 - x^2)^2 + 36x^2y^2}$$

$$= \sqrt{9x^4 + 9y^4 - 18x^2y^2 + 36x^2y^2}$$

$$= \sqrt{9x^4 + 9y^4 + 18x^2y^2}$$

$$= 3\sqrt{x^4 + y^2 + 2x^2y^2}$$

$$= 3(x^2 + y^2)$$

$$\text{Direction} = \theta = \tan^{-1}\left[\frac{6xy}{3(y^2 - x^2)}\right]$$

$$= \tan^{-1}\left[\frac{2xy}{y^2 - x^2}\right]$$

Ex. 4.5: A flow field is characterized by the stream function $\psi = 3x^2y - y^3$. Check whether this is a possible case of fluid flow. Is this flow irrotational? If not, calculate the velocity. Show that the magnitude of velocity at any point in the flow field depends only on its distance from the origin.

Solution. :

To find out velocity components,

	$u = -\dfrac{\partial\psi}{\partial y} \qquad v = \dfrac{\partial\psi}{\partial x}$ $u = -3x^2 + 3y^2 \qquad v = 6xy$ To check whether the flow is continuous, $\dfrac{\partial u}{\partial x} + \dfrac{\partial v}{\partial y} = 0$ $-6x + 6x = 0$ ∴ The flow is continuous. To decide whether flow is irrotational. $\omega_z = \dfrac{\partial v}{\partial x} - \dfrac{\partial u}{\partial y}$ $\therefore \omega_z = \dfrac{1}{2}(6y - 6y) = 0$ ∴ The flow is irrotational. For Magnitude of velocity, $\lvert V \rvert = \sqrt{u^2 + v^2}$ $= \sqrt{+9(x^2 - y^2)^2 + 36x^2y^2}$ $= \sqrt{9x^4 - 18x^2y^2 + 9y^4 + 36x^2y^2}$ $= \sqrt{9x^4 + 18x^2y^2 + 9y^4}$ $\lvert V \rvert = 3(x^2 + y^2)$ It proves that magnitude of velocity depends on distance from origin.

Ex. 4.6 :	In a two dimensional fluid motion the stream function is given by $\psi = 4xy$. 1 Sketch the stream lines. 2 Determine the potential function and sketch equipotential lines.

3 Determine the velocity at point (3,1).
4 Determine the convective acceleration at point (3,1).

Solution. :

$$\frac{\partial \psi}{\partial y} = -u \qquad \frac{\partial \psi}{\partial x} = v$$

$$v = 4y \qquad u = -4x$$

For Velocity at (3,1)

$$u = -12 \qquad v = 4$$

$$|V\_| = \sqrt{u^2 + v^2} = \sqrt{144 + 16} = 12.64$$

$$\tan\theta = -\frac{4}{12}$$

$$\theta = -\tan^{-1}(0.33) = 18°26'$$

$$\frac{\partial \phi}{\partial x} = -u = 4x$$

Integrating w.r.t. x

$$\phi = 2x^2 + f(y)$$

$$\therefore \quad \frac{\partial \phi}{\partial y} = f'(y)$$

But $\frac{\partial \phi}{\partial y} = -v$ which is $-4y$

$$\therefore \quad f'(y) = -4y$$

$$f(y) = -2y^2 + c$$

Substituting value of $f(y)$ in Equation (1)

$$\phi = 2x^2 - 2y^2 + c$$

$$\phi = 2(x^2 - y^2) + c$$

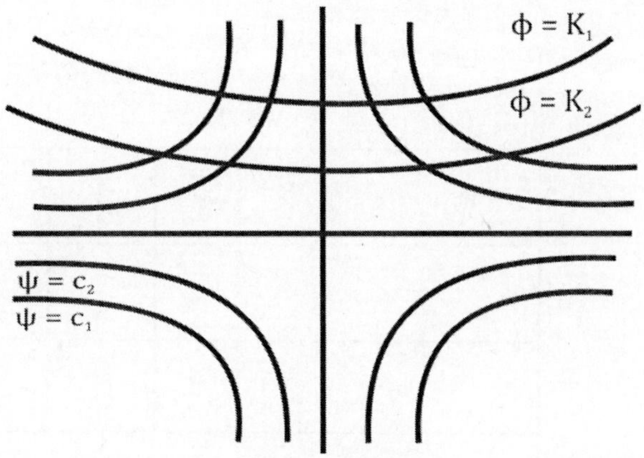

Fig. Ex.4.6

$$a_x = u\frac{\partial u}{\partial x} + v\frac{\partial u}{\partial y} = (-4x)(-4) + 4y(0) = +16x$$

$$a_y = u\frac{\partial v}{\partial x} + v\frac{\partial v}{\partial y} = (-4x)(0) + (4y)(4) = 16y$$

For point (3,1) $(a_x, a_y) = (48, 16)$ $\therefore |a| = 50.59$

Ex. 4.7 : The stream function for two dimensional flow is given by $\psi = 2xy$ in the range of values of x and y between 0 to 5. Plot stream lines passing through the points (1,1), (1,2), (2,2). The drawing need not be to the scale. Determine the velocity in magnitude and direction at (1,2).

Solution :

$$u = -\frac{\partial \psi}{\partial y} = -2x;\ v = \frac{\partial \psi}{\partial x} = 2y$$

$$\therefore |V| \text{ for } (1,2) = \sqrt{u^2 + v^2} = \sqrt{4 + 16} = \sqrt{20} = 4.472 \text{ units}$$

$\tan\theta = \frac{v}{u} = -2 \;\therefore\; \tan{-1}(-2) = -63.43$ (In second quadrant)

$= 116.566^0$

Now, x	1	2	3	4	5	1
y	1	2	3	4	5	2
ψ	2	8	18	32	50	4

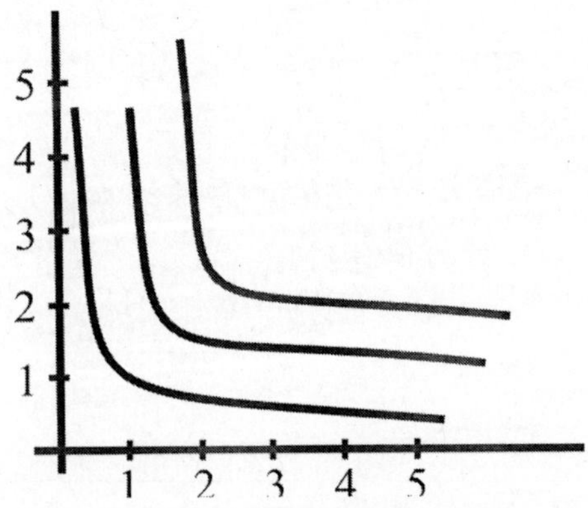

Fig. Ex. 4.7 Sketch of stream lines

Ex. 4.8 : For an incompressible flow represented by $\psi = x^2 - y^2$, calculate the total acceleration vector and show that it is proportional to the radius vector.

Solution :

$$u = -\frac{\partial\psi}{\partial y} = 2y;\; v = \frac{\partial\psi}{\partial x} = 2x$$

	$\frac{\partial u}{\partial x} = 0;\ \frac{\partial u}{\partial y} = 2;\ \frac{\partial v}{\partial x} = 2;\ \frac{\partial v}{\partial y} = 0$ $a = \sqrt{a_x^2 + a_y^2}$ where $a_x = u\frac{\partial u}{\partial x} + v\frac{\partial u}{\partial y}$ and $ay = u\frac{\partial v}{\partial x} + v\frac{\partial v}{\partial y}$ $\therefore\quad a_x = 0 + 2x\,(2) = 4x$ $a_y = 2(2y) + 0 = 4y$ and $\therefore\quad a = \sqrt{a_x^2 + a_y^2} = 4\sqrt{x^2 + y^2}$ $a = 4\sqrt{x^2 + y^2} = 4r$ Since acceleration a = 4r, it is proportional to radius r.

Ex. 4.9:	If $\phi = 3xy$, find x and y components of velocity at (1,3) and (3,3). Determine the discharge passing between stream lines passing through these points. **Solution :** $u = \frac{\partial \phi}{\partial x};\qquad v = -\frac{\partial \phi}{\partial y}$ $\therefore\qquad u = -3y;\qquad v = -3x$ Velocity component for (1,3) $= (u, v)_{1,3} = (9, -3)$

	Velocity component for (3,3) $= (u, v)_{3,3} = (-9, -9)$ Total derivative ψ can be written as, $d\psi = \frac{\partial \psi}{\partial x} dx + \frac{\partial \psi}{\partial y} dy$ but $\frac{\partial \psi}{\partial x} = v; \quad \frac{\partial \psi}{\partial x} = -u$ $\therefore d\psi = vdx - udy = -3xdx + 3ydy$ Integrating, $\psi = -\frac{3x^2}{2} + \frac{3y^2}{2} + c$ Where c is constant of integration. Discharge between the stream lines through (1,3) and (3,3). $\psi_{(1,3)} - \psi_{(3,3)} = \left(-\frac{3}{2} + \frac{27}{2}\right) - \left(-\frac{27}{2} + \frac{27}{2}\right) = 12$ units

Ex. 4.10:	Velocity components of a flow are given by $u = -x$, $v = 2y$, $w = 5 - z$. Derive the equation of stream line passing through point (2,1,1). **Solution :** $\frac{dx}{u} = \frac{dy}{v} = \frac{dz}{w}$ $\frac{dx}{-x} = \frac{dy}{2y} = \frac{dz}{5-z}$ $\frac{dx}{-x} = \frac{dy}{2y}$

	Integrate
	$$-\log x = \frac{1}{2}\log y + C_1$$
	$$-\left[\log x + \frac{1}{2}\log y\right] = c_1$$
	$$-\log x\sqrt{y} = c_2 \quad \therefore x\sqrt{y} = c_3$$
at (2,1,1)	$$c_3 = 2$$
	$$\therefore \quad x\sqrt{y} = c$$
	Now
	$$-\frac{dx}{x} = \frac{dz}{5-z}$$
	$$-\log x = -\log(5-z) + c_4$$
	$$\log\left(\frac{5-z}{x}\right) = c_5$$
	$$\therefore \quad \left(\frac{5-z}{x}\right) = c_6$$
at (2,1,1)	$$\left(\frac{5-z}{x}\right) = 2$$
	$$\therefore \quad x\sqrt{y} = \left(\frac{5-z}{x}\right) = 2$$

Ex. 4.11:	Determine the stream function if velocity components of a 2-D incompressible fluid flow are given by $u = \frac{y^3}{3} + 2x - x^2y$ and $v = xy^2 - 2y - \frac{x^3}{3}$ **Solutions :**

	$u = \frac{y^3}{3} + 2x - x^2y = -\frac{\partial\psi}{\partial y}$ by definition. $\therefore$ Integrating w.r.t. y $\psi = -\frac{y^4}{12} - 2xy + \frac{x^2y^2}{2} + f(x)$ Differentiating w.r.t. x $\frac{\partial\psi}{\partial y} = -2y + x^2y + f'(x)$ But $\frac{\partial\psi}{\partial x} = V$ $\therefore \frac{\partial\psi}{\partial x} = x^2y - 2y - \frac{x^2}{3}$ ($\therefore$ by given data) Comparing Equation (1) and (2) $-2y + x^2y + f'(x) = x^2y - 2y - \frac{x^2}{3}$ $\therefore f'(x) = -\frac{x^2}{3}$ $\therefore f(x) = -\frac{x^3}{9} + c$ Substituting value of $f(x)$ $\psi = -\frac{y^4}{12} + \frac{x^2y^2}{2} - 2xy - \frac{x^3}{9} + c$

Ex. 4.12 :	A uniform steady incompressible flow field has a horizontal component of velocity 4 m/s and a vertical component of velocity 3 m/s. Determine expressions for velocity potential and stream function. Sketch the lines of velocity potential and stream function.

Solution :

$$\frac{\partial \phi}{\partial x} = -u = -4$$

Integrating w.r.t. x $\qquad \phi = -4x + f(y)$

Differentiating w.r.t. y $\qquad \frac{\partial \phi}{\partial x} = f'(y)$

But by definition $\qquad \frac{\partial \phi}{\partial y} = -v$

$$\frac{\partial \phi}{\partial y} = -3$$

Comparing values of $\frac{\partial \phi}{\partial y}$ from steps 2 and 3.

$$f'(y) = -3$$

$$\therefore \quad f(y) = -3y + c$$

$$\therefore \quad \phi = -4x - 3y + c$$

Now $\qquad \frac{\partial \psi}{\partial x} = v = 3$

Integrating w.r.t. x $\qquad \psi = 3x + f(y)$

Differentiating w.r.t. y, $\qquad \frac{\partial \psi}{\partial y} = f'(y)$

But $\qquad \frac{\partial \psi}{\partial y} = -u = -4 \qquad$ by definition.

$$f'(y) = -4$$

$$\therefore \quad f(y) = -4y + c$$

$$\therefore \quad \psi = 3x - 4y + c$$

$\phi = -4x - 3y + c$ is a straight line passing through (0,0) and slope $-\frac{4}{3}$

$\psi = 3x - 4y + c$ is also a straight line passing through (0,0) and slope $\frac{3}{4}$

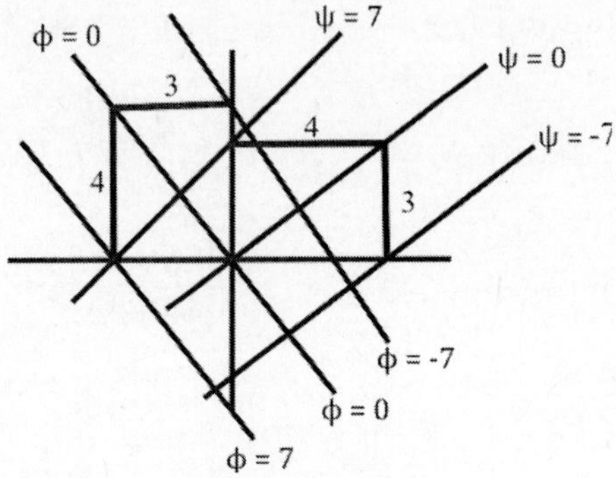

Fig. Ex.4.12

Ex. 4.13: Determine whether following specified flows are rotational or otherwise. Determine the expression for velocity potential in case of irrotational flow.

(i) $u = y,\qquad v = -\frac{3}{2}x$

(ii) $u = xy^2\qquad v = x^2y$

Solution:

(i) $$\frac{\partial v}{\partial x} = -\frac{3}{2},\qquad \frac{\partial u}{\partial y} = 1,\qquad \frac{\partial v}{\partial x} \neq \frac{\partial u}{\partial y}$$

$\therefore\ \omega z \neq 0\qquad \therefore$ Flow is rotational

$$\frac{\partial v}{\partial x} = 2xy,\qquad \frac{\partial u}{\partial y} = 2xy,\qquad \frac{\partial v}{\partial x} \neq \frac{\partial u}{\partial y}$$

$\therefore\ \omega z = 0\qquad \therefore$ Flow is irrotational

For (ii) to find out ϕ; by definition.

	$$\frac{\partial \phi}{\partial x} = -u = -xy^2$$ Integrating w.r.t. x $$\phi = \frac{-x^2y^2}{2} + f(y)$$ Differentiating w.r.t. y $$\frac{\partial \phi}{\partial x} = -x^2y + f'(y)$$ But $\frac{\partial \phi}{\partial y} = -v$ by definition. $$\therefore \quad \frac{\partial \phi}{\partial y} = -x^2y$$ Comparing two values of $\frac{\partial \phi}{\partial y}$ $$-x^2y + f'(y) = -x^2y$$ $$f'(y) = 0$$ $$f(y) = \text{constant}$$ $$\therefore \quad \phi = -\frac{x^2y^2}{2} + c$$

Ex. 4.14:	The velocity vector in a fluid flow is given by $$V = (4x^3)i - (10x^2y)j + (2t)k$$ Obtain expression for velocity vector and acceleration vector at a point (2,1,3) at time $t = 1$sec. Also calculate the value of velocity and acceleration at the given point. **Solution :**

	$u = 4x^3, \quad v = -10x^2y, \quad w = 2t$ at point (2,1,3) $\quad \lvert V \rvert = \sqrt{32 - 40 + 2} = 51.26$ units $a_x = u\frac{\partial u}{\partial x} + v\frac{\partial u}{\partial y} + w\frac{\partial u}{\partial z} + \frac{\partial u}{\partial t}$ $\frac{\partial u}{\partial x} = 12x^2, \quad \frac{\partial u}{\partial y} = \frac{\partial w}{\partial z} = 0$ $a_x = (32)(48) + (-40)(0) + 2(0) + 0$ $a_x = 1536$ units $a_y = u\frac{\partial v}{\partial x} + v\frac{\partial v}{\partial y} + w\frac{\partial v}{\partial z} + \frac{\partial v}{\partial t}$ $\frac{\partial v}{\partial x} = -20xy, \quad \frac{\partial v}{\partial y} = -10x^2, \quad \frac{\partial w}{\partial z} = 0$ $32 \times (-40) + (-40)(-40) + 0 = 0$ $a_y = 320$ units $a_z = u\frac{\partial w}{\partial x} + v\frac{\partial w}{\partial y} + w\frac{\partial w}{\partial z} + \frac{\partial w}{\partial t}$ $\frac{\partial w}{\partial t} = 2, \quad \frac{\partial w}{\partial x} = \frac{\partial w}{\partial y} = \frac{\partial w}{\partial z} = 0$ $a_z = 2$ units $a_a = 1536i + 320j + 2k$ $\lvert a \rvert = \sqrt{1536^2 + 320^2 + 2^2} = 1568.98$ units

Ex. 4.15:	A fluid has following flow field. (1) $C = \frac{5x^3}{i} - (15x^2y)j + T$ km/s

	Obtain expressions for velocity and acceleration vector at a point (1,2,3) at time T = 1 sec. Also calculate the value of velocity and acceleration at a given point. **Solution :** $u = 5x^3,\ \ v = -15x^2y,\ \ w = T$ $\frac{\partial u}{\partial x} = 15x^2, \qquad \frac{\partial u}{\partial y} = 0 = \frac{\partial u}{\partial z} = \frac{\partial u}{\partial t} = 0$ $\frac{\partial v}{\partial x} = -30xy, \quad \frac{\partial v}{\partial y} = 15x^2, \quad \frac{\partial v}{\partial z} = 0, \quad \frac{\partial v}{\partial t} = 0$ $\frac{\partial w}{\partial x} = \frac{\partial w}{\partial y} = \frac{\partial w}{\partial z} = 0, \quad \frac{\partial w}{\partial t} = 1$ at (1,2,3), $u = 5,\ \ v = -30,\ \ w = 1$ $\therefore\ \ \lvert V \rvert = \sqrt{25 + 900 + 1}$ $= 30.43$ units m/s $a_x = u\frac{\partial u}{\partial x} + v\frac{\partial u}{\partial y} + w\frac{\partial u}{\partial y} + \frac{\partial u}{\partial t}$ $= 5(15)$ $a_x = 75$ $a_y = u\frac{\partial v}{\partial x} + v\frac{\partial v}{\partial y} + w\frac{\partial v}{\partial z} + w\frac{\partial v}{\partial z} + \frac{\partial v}{\partial t}$ $= 5(-60) + (-30)(-15) + 0 + 0$ $= 150$ $a_z = u\frac{\partial w}{\partial x} + v\frac{\partial w}{\partial y} + w\frac{\partial w}{\partial z} + \frac{\partial w}{\partial t}$

	$= 1$ $\lvert a \rvert = \sqrt{75^2 + 150^2 + 1^2}$ $= 167.71\ m/s^2$

UNIT SUMMARY

Unit is summarized in the following points :

1 If the fluid characteristics like velocity, pressure density etc. do not change at a point with respect to time, the fluid flow is called as steady flow. If they change with respect to time, the fluid flow is called as unsteady flow.

$\left(\frac{\partial v}{\partial t}\right)_{x_0,y_0,z_0} = 0$ for steady flow. $\left(\frac{\partial v}{\partial t}\right)_{x_0,y_0,z_0} \neq 0$ for unsteady flow

2 If the velocity in a fluid flow does not change with respect to space, the flow is said to be uniform flow otherwise non- uniform flow.

$\left(\frac{\partial V}{\partial s}\right)_{t=t_1} = 0$ for uniform flow. $\left(\frac{\partial V}{\partial s}\right)_{t=t_1} \neq 0$ for non-uniform flow

3 If Reynolds number in a pipe is less than 2000 the flow is said to be laminar, if it is more than 4000 the flow is turbulent.

4 Differential equation of a stream line is $udy - vdx = 0$.

5 For steady flow, path lines, stream lines and streak lines coincide with each other.

6 Continuity equation in differential form is, $\frac{\partial u}{\partial x} + \frac{\partial v}{\partial y} = 0$ for a two dimensional flow.

7 Continuity equation can also be written as $Q = A_1V_1 = A_2V_2 =$ constant.

8 The components of acceleration in x, y, and z directions are

$$a_x = u\frac{\partial u}{\partial x} + v\frac{\partial u}{\partial y} + w\frac{\partial u}{\partial z} + \frac{\partial u}{\partial t}$$

$$a_y = u\frac{\partial v}{\partial x} + v\frac{\partial v}{\partial y} + w\frac{\partial v}{\partial z} + \frac{\partial v}{\partial t}$$

$$a_z = u\frac{\partial w}{\partial x} + v\frac{\partial w}{\partial y} + w\frac{\partial w}{\partial z} + \frac{\partial w}{\partial t}$$

9 The tangential and normal components of acceleration are :

$$a_s = V_s\frac{\partial V_s}{\partial s} + \frac{\partial V_s}{\partial t}$$

$$a_n = \frac{\partial V_n}{\partial t} + \frac{V_s^2}{R}$$

10 Angular deformation or shear strain rate is given by,

$$\text{Shear strain rate} = \frac{1}{2}\left(\frac{\partial v}{\partial x} + \frac{\partial u}{\partial y}\right)$$

11 Rotational components of a fluid particle are :

$$\omega x = \frac{1}{2}\left(\frac{\partial v}{\partial x} - \frac{\partial u}{\partial y}\right); \quad \omega y = \frac{1}{2}\left(\frac{\partial w}{\partial y} - \frac{\partial v}{\partial z}\right); \quad \omega z = \frac{1}{2}\left(\frac{\partial v}{\partial x} - \frac{\partial u}{\partial y}\right)$$

12 The components of velocity in x, y and z directions in terms of velocity potential (ϕ) are

$$u = -\frac{\partial \phi}{\partial x}, \quad v = -\frac{\partial \phi}{\partial y}, \quad w = -\frac{\partial \phi}{\partial z}$$

13 The stream function (ψ) is defined only for two-dimensional flow. The velocity components in x and y directions in terms of stream function are $u = -\frac{\partial \psi}{\partial y}$ and

$$v = \frac{\partial \psi}{\partial x}$$

4.11: Exercise

4.11.1: Objective Questions:

1. A flow in which each fluid particles have definite paths and their paths do not cross each other, is called
   (a) Steady flow
   (b) Uniform flow
   (c) Streamline flow
   (d) Turbulent flow

   Ans: (c)

2. Laminar flow usually occurs at ______ velocity.
   (a) low
   (b) high
   (c) very high
   (d) sometimes high and sometimes low

   Ans: (a)

3. The continuity equation is the result of application of the following law to the flow field
   (a) First law of thermodynamics
   (b) Conservation of energy
   (*c*) Newton's second law of motion
   (*d*) Conservation of mass.

   Ans: (d)

4. A path line describes
   (a) The velocity direction at all points on the line
   (b) The path followed by particles in a flow
   (c) The path over a period of times of a single particle that has passed out at a point
   (*d*) The instantaneous position of all particles that have passed a point.

   Ans: (c)

5. The stream function is
   (a) constant along an equipotential line
   (b) along a stream line
   (c) defined only in irrotational flow
   (d) defined only for incompressible flow.
   Ans: (b)

6. A potential function
   (*a*) is constant along a stream line
   (*b*) is definable if a stream function is available for the flow
   (*c*) describes the flow if it is rotational
   (*d*) describes the flow if it is irrotational.
   Ans: (d)

7. A flow is defined by $u = 2\ (1 + t)$, $v = 3(1 + t)$ where t is the time. The velocity at $t = 2$ is,
   (a) 6 (b) 9
   (*c*) 10.82 (*d*) 6.7.
   Ans: (c)

8. For every potential function a stream function should exist.
   (a) Correct (b) Incorrect
   Ans: (a)

9. Stream function can exist only for irrotational flow.
   (a) Correct (b) Incorrect
   Ans: (b)

10. The stream lines and equipotential lines for a flow field are
(a) Parallel to each other
(b) Orthogonal to each other
(c) Inclined with each other
(d) Parabolic

Ans: (b)

11. The condition to satisfied by irrotational flow is

(a) $u = \frac{\partial \phi}{\partial y}$ $v = -\frac{\partial \phi}{\partial y}$ $u = \partial ø/\partial y$ $v = -\partial ø/\partial y$

(b) $u = -\partial ø/\partial x,\ = -\partial ø/\partial y$

(c) $\partial v/\partial x = \partial u/\partial y$

(d) $ø = øA + øB.$

Ans: (b)

12. The flow through an expanding pipe at a constant rate is called as
(a) Steady uniform flow
(b) Unsteady uniform flow
(c) Steady non uniform flow
(d) Unsteady non uniform flow

Ans: (c)

13. For a flow the velocity vector is expressed as V = 3xi -3yj, then the equation of the streamline passing through the point (1,1) is
(a) $xy = 1$
(b) $x^2y = 1$
(c) $x^2y^2 = 1$
(d) $xy^2 = 1$

Ans: (a)

14. Flow of liquid under pressure through long pipe lines of varying diameter is
    (a) Steady flow
    (b) Unsteady flow
    (c) Uniform flow
    (d) Non uniform flow

    Ans: (d)

15. Flow of liquid through a tapering pipe at varying rate is
    (a) Steady uniform flow
    (b) Unsteady uniform flow
    (c) Steady non uniform flow
    (d) Unsteady non uniform flow

    Ans: (d)

4.11.2: Theory Questions :

Q.1 Explain Eularian method of representing fluid motion.

Q.2 What do you understand by Kinematics of fluid flow ?

Q.3 Which are the methods of analysis of fluid flow ?

Q.4 What is the difference between Lagrangian and Eulerian methods of studying a fluid flow ?

Q.5 Define : (i) Path line, (ii) Stream line, (iii) Stream tube and (iv) Streak line. What is the special feature of concept of stream tube?

Q.6 Derive equation for stream line, $udy - vdx = 0$, for a plane flow in x - y plane.

Q.7 Describe the types of flow bringing out their characteristics.

Q.8 Define the following terms :
(i) Discharge
(ii) Mean velocity
(iii) Stream line
(iv) Unsteady flow.

Q.9 How are the flows classified? Give their governing conditions.

Q.10 Define :

(i) Steady flow and Unsteady flow

(ii) Uniform and non-uniform flow

What combinations of above flows are possible? Give one example of each such combination.

Q.11 Distinguish between Rotational and Irrotational flow.

Q.12 Define 1 - D, 2 - D and 3 - D flows and give one example of each.

Q.13 Prove that in a two dimensional flow field, rotation of the element is given by the expression.

$$\omega = \frac{1}{2}\left(\frac{\partial v}{\partial x} - \frac{\partial u}{\partial y}\right)$$

Q.14 Prove that potential flow is also irrotational flow.

Q.15 Derive general form of continuity equation for two dimensional flow.

Q.16 Derive continuity equation $\frac{\partial u}{\partial x} + \frac{\partial v}{\partial y} + \frac{\partial w}{\partial z} = 0$ with usual notations.

Q.17 Derive the continuity equation in one dimensional flow.

Q.18 Explain the concept of stream function and velocity potential. Establish the condition that the lines of constant stream function and velocity potential are mutually perpendicular.

Q.19 What is a flow net? What are its uses? What are the methods of drawing a flow net?

Q.20 Explain electrical analogy method of drawing a flow net.

Q.21 Distinguish clearly between Laminar flow and Turbulent flow.

Q.22 Classify the following flows as steady, unsteady, uniform and non-uniform.

(i) Constant discharge through converging pipe under constant temperature.

(ii) Flow of oil in Redwood Viscometer experiment.

(iii) Flow of constant discharge in a long rectangular passage of constant width of wind tunnel.

(iv) Constant discharge along a long bend.

Q.23 What do you understand by :

(i) Total acceleration,

(ii) Local acceleration and

(iii) Convective acceleration

(May 2000, 3 Marks)

Expression (4.23), (4.24) and (4.25) represent the three components of total acceleration. The terms $\frac{\partial u}{\partial t}, \frac{\partial v}{\partial t}, \frac{\partial w}{\partial t}$ represent local acceleration and remaining terms represent convective acceleration.

Q.24 Show that stream lines and equipotential lines are orthogonal to each other.

Q.25 A stream function is given by $\psi = 3xy$

Determine :

(i) Whether flow is possible

(ii) Whether flow is rotational or irrotational

(iii) The potential function ϕ

(iv) Acceleration component at (1,1).

Q.26 Prove that in a two dimensional flow field rotation of the element is given by the expression.

Q.27 Derive three dimensional continuity equation.

Q.28 Differentiate between rotational and irrotational flow.

4.11.3: Problems:

1. State whether the flow of liquid given by u = $4x$ and v = -4y is

(i) continuous (ii) irrotational

(Ans: the flow is continuous as well as irrotational)

2. For the above mentioned flow (u = $4x$ and v = -4y), determine the stream function if the flow is possible.

   (Ans: the flow is possible. $\psi = -2x^2 - 2y^2 + C$)

3. Determine whether the following specified flows are rotational or otherwise. Determine velocity potential (1) $u = y$, $v = (-\frac{3}{2})x$ (2) $u = xy^2$, $v = x^2y$

   (Ans: (1) the flow is rotational, Velocity potential does not exist,

   (2) the flow is irrotational and $\emptyset = \frac{-x^2y^2}{2} + C$)

4. Velocity vector in a 2-D flow field is given by $\bar{V} = x^2yi - xy^2j$ check whether flow is possible. Also find whether the flow is irrotational. If so find magnitude of rotation at a point (1,1).

   (Ans : Flow is possible and rotational. $wz = -1$ unit)

5. For a two-dimensional potential flow, velocity potential $\phi = x(2y - 1)$ find (b) whether flow is possible and expression for stream function.

   (Ans : flow is continuous, $\psi = -x^2 + y^2 - y + c$)

QR CODES FOR SUPPORTING VIDEO LINKS

(1) (2) (3) (4) (5)

REFERENCES AND SUGGESTED READINGS are same as earlier unit

5 FLUID DYNAMICS

UNIT SPECIFICS

Through this unit we have discussed the following aspects:

- *Forces acting on fluid in motion*
- *Euler's equation, Bernoulli's equation and Momentum equation*
- *Working of Venturimeter, orifice meter and pitot tube and use of these equipment to measure discharge and velocity*
- *Forces exerted by fluid flow on pipe bend*
- *Experimental determination of coefficient of discharge of Venturimeter and orifice meter*

This is followed by large number of solved examples. The students are encouraged to solve the objective questions, long answer questions and numerical problems to judge ones understanding. The practical on determination of coefficient of discharge of Venturimeter and orifice meter is included followed by a list of references for additional reading.

RATIONALE

This unit introduces concept of fluid dynamics in which forces responsible for motion are considered for analysis. The Euler's equation followed by Bernoulli's are then derived which form the working principle of many discharge and velocity equipment. Working of the Venturimeter, Orificemeter and pitot tube is therefore discussed followed by applying Bernoulli's equation to this equipment. The momentum equation needs to be discussed when one is studying fluid dynamics owing to its many applications in fluid dynamics

PRE-REQUISITES

Mathematics: Derivatives (Class XII)

Physics: Mechanics (Class XII)

Fluid Mechanics: Unit I and II

UNIT OUTCOMES

List of outcomes of this unit is as follows:

(At the end of this unit, students will understand..)

U5-01: Forces acting on fluid in motion

U5-02: Euler's equation

U5-03: Bernoulli's equation

U5-04: Momentum equation

U5-05 Working of Venturimeter, orifice meter and pitot tube

U5-06 Experimental procedure to determine coefficient of discharge of Venturimeter and orifice meter

Unit-5 Outcomes	*EXPECTED MAPPING WITH COURSE OUTCOMES (1-WeakCorrelation;2-Mediumcorrelation;3-StrongCorrelation)*					
	CO-1	*CO-2*	*CO-3*	*CO-4*	*CO-5*	*CO-6*
U5-O1	3	-	-	-	-	-
U5-O2	3	2	-	-	-	-
U5-O3	3	2	-	-	-	-
U5-O4	3	2	-	3	-	-
U5-O5	3	-	-	3	-	-
U5-O6	3	-	-	-	-	-

5.1 Introduction:

In kinematics the space-time relationships of the fluid motion have been discussed without considering the 'forces' responsible for the motion. In the present unit the forces responsible for fluid motion are considered to determine the resulting accelerations and the energy change involved in the flow phenomenon. This aspect of fluid motion is known as the Dynamics of fluid flow. Similar to the mechanics of solids, the mechanics of fluids is governed by Newton's second law of motion which states that Force = Mass × Acceleration. The fluid is assumed to be incompressible and non-viscous.

5.2 Surface and Body Forces:

According to Newton's second law of motion, the net force F_x acting on a fluid element in the X direction is equal to mass m of the fluid element multiplied by the acceleration a_x in the x-direction. Thus,

$$F_x = m.a_x \tag{5.1}$$

In the fluid flow various forces influence the fluid motion namely force due to gravity (F_g), pressure (F_p), viscosity (F_v), surface tension (F_s) compressibility (F_e) and turbulence (F_t). Out of these some are body forces and some are surface forces. By body forces we mean the forces which act on the body like weight of the body, pressure acting on the body. While the forces which are acting along the body surface like shear forces or surface tension are termed as surface forces. It is to be understood that all the forces may not act on the fluid at the same time as well as all are not equally dominant as discussed in the next section.

5.3 Equations of Motion :

Thus, as mentioned in the earlier section, in Equation (5.1) the net force along X axis,

$$F_x = (F_g)_x + (F_P)_x + (F_v)_x + (F_s)_x + (F_e)_x + (F_t)_x \tag{5.2}$$

In case of incompressible fluids, the force due to compressibility, F_e is negligible, then the resulting force is,

$$F_x = (F_g)_x + (F_P)_x + (F_v)_x + (F_t)_x \tag{5.3}$$

and equation of motions are called Reynold's equation of motion.

For flow where force due to turbulence Ft is negligible, the resulting equations of motion are known as Navier-Stokes Equation

$$F_x = (F_g)_x + (F_P)_x + (F_v)_x \quad (5.4)$$

If the flow is assumed to be ideal, viscous force F_V is zero and equation of motion are known as Euler's equation of motion which will have the force due to gravity and pressure acting on the fluid

$$F_x = (F_g)_x + (F_P)_x \quad (5.5)$$

Considering the first course on Fluid Mechanics, the present text deals with Euler's equation of motion only and thus other equations are out of scope of the present syllabus.

5.4 Euler's Equation of Motion:

The forces acting on fluid in motion due to gravity and pressure are taken into consideration. This is derived by considering the motion of a fluid element along a streamline.

In figure 5.1, a stream tube of fluid of length ds and area dA is isolated as a free body from the moving fluid. The external forces are due to the pressure acting on the tube and gravitational force (weight of the stream tube).

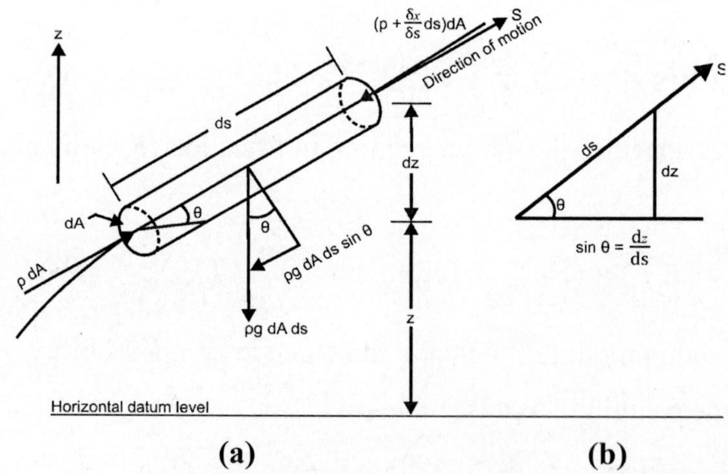

Fig. 5.1

The length ds is so small that curvature of the streamlines over this distance may be neglected. The external forces which act on the chosen fluid element causing the acceleration of flow are as follows:

1 The resultant force due to pressure acting on the end surfaces in the direction of flow is,

$$p\,dA - \left(p + \frac{\partial p}{\partial s} ds\right) dA = -\frac{\partial p}{\partial s} ds\, dA \tag{5.6}$$

2 The component of gravity force of the fluid element in the direction of motion is,

$$-\rho\, g\, dA\, ds \sin\theta = -\rho\, g\, dA\, ds \frac{dz}{ds} \quad \text{(refer Fig 5.1b)} \tag{5.7}$$

The resultant external force acting on the free body in the direction of flow is sum of the pressure force and gravitational force

$$\sum Fs = -\frac{\partial p}{\partial s} ds\, dA - \rho\, g\, dA\, ds \frac{dz}{ds} \tag{5.8}$$

The sum of these forces is equal to mass multiplied by acceleration (Newton's second law of motion).

$$\text{mass} = m\, \rho\, dA\, ds$$

$$a_s = \frac{\partial V}{\partial t} = \frac{\partial S}{\partial t} \cdot \frac{\partial V}{\partial s} = V \frac{\partial V}{\partial s}$$

$\therefore$

$$m\, a_s = \rho\, dA\, ds \cdot V \frac{\partial V}{\partial s} \tag{5.9}$$

According to Newton's law of motion, using Equations (5.8) and (5.9)

$$-\frac{\partial p}{\partial s} ds\, dA - \rho\, g\, dA\, ds \frac{dz}{ds} = \rho\, dA\, ds\, V \frac{\partial V}{\partial s} \tag{5.10}$$

or

Dividing by dA ds

$$-\frac{1}{\rho} \frac{\partial p}{\partial s} - g \frac{dz}{ds} = V \frac{\partial V}{\partial s} \tag{5.11}$$

or $$\frac{1}{\rho}\frac{\partial p}{\partial s} + g\frac{dz}{ds} + V\frac{\partial V}{\partial s} = 0 \tag{5.12}$$

The equation (5.12) is known as Euler's equation of motion which is written in differential form. The same can be written in a common form as –

$$\frac{\partial p}{\rho} + g\,dz + V\,dv = 0 \tag{5.13}$$

5.5 Bernoulli's Equation:

Integration of Euler's Equation Along a Streamline for Steady Flow:

The Equation (5.13) cannot he integrated completely with respect to distance 's' unless mass density of fluid 'ρ' is either constant or a known function of pressure 'p'. For fluid of constant density (incompressible), the result of integration is,

$$\frac{P}{\rho} + \frac{V^2}{2} + g\,z = \text{constant} \tag{5.14}$$

This is known as Energy equation or Bernoulli's equation. The Bernoulli's equation relates pressure changes to velocity and elevation changes along a streamline. However, it gives correct results only when applied to a flow situation where the following assumptions made are reasonable.

1 Only pressure and gravity forces are to be considered.

2 Flow is along the streamline.

3 Flow is steady.

4 Flow is with negligible viscosity.

5 Flow is ideal.

6 Fluid is incompressible.

7 Velocity is uniform over the section.

Alternatively,

$$\frac{P}{\rho g} + \frac{V^2}{2g} + z = \text{constant} = \text{say H} \tag{5.15}$$

In the Equation (5.15 each term has dimensions of energy per unit weight; such as Nm/N or more simply as 'm', indicating a scalar quantity. It is thus convenient to refer to each term as a 'head'. Therefore, H is called the total energy head, and is the sum of pressure, velocity and potential heads. The statement of Bernoulli's theorem cab be stated as for steady flow of an incompressible frictionless fluid total energy at any point remains constant.

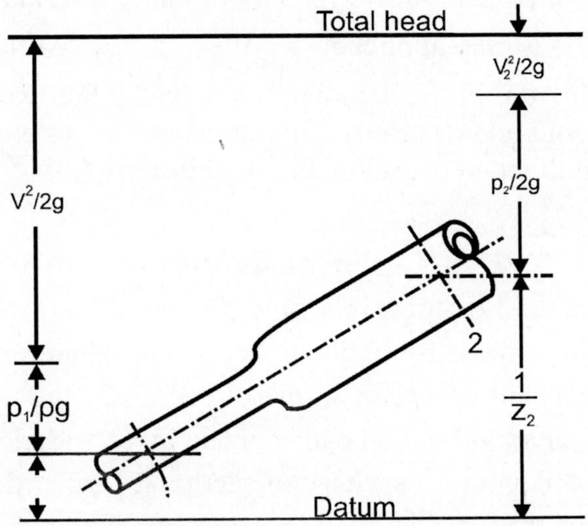

Fig. 5.2 Total energy heads

Figure (5.2) depicts the relationship among the types of energy. It is to be noted that because of the assumption that no energy is lost or added, the total head remains constant and height of each head term varies as predicted by Bernoulli's equation.

The Equation (5.15) can be applied for a single streamline. The sum of the three terms is constant along any streamline, but the value of the constant may be different for different streamlines in a given stream. If the equation is integrated along the streamline between any two points indicated by suffixes 1 and 2.

$$\frac{p_1}{\rho g}+\frac{v_1^2}{2g}+Z_1=\frac{p_2}{\rho g}+\frac{v_2^2}{2g}+Z_2 \qquad (5.16)$$

5.5.1 Limitations on Bernoulli's Equation:

Although Bernoulli's equation is applicable to a large number of practical problems, there are several limitations that must be understood in order to apply it properly.

1 There can be no energy lost due to friction.

2 There can be no heat transferred into or out of the fluid.

3 The equation is only applicable for flow of incompressible fluids.

4 There can be no mechanical devices between the two sections of interest which would add or remove energy from the system, since the equation states that the total energy in the fluid is constant.

5.5.2 Modification to Bernoulli's Equation:

It is evident that in the flow of a real fluid, there will be losses of energy due to friction, separation and formation of eddies etc. Therefore, the total head will not be constant but decreases in the direction of flow as a result of energy dissipation. The modification can be made by considering that the total head at section 1 is equal to total head at section 2 further downstream, plus the losses between the two sections. Thus,

$$\frac{P_1}{\rho g}+\frac{V_1^2}{2g}+Z_1 = \frac{P_2}{\rho g}+\frac{V_2^2}{2g}+Z_2 + \text{losses} \tag{5.17}$$

In which the losses are also 'Nm/N' or in 'm' head of the fluid concerned. A further modification can be made to take into account an addition or subtraction of energy between the two sections.

Pumps, which convert mechanical energy into hydraulic energy and turbines which perform the reverse function are typical examples of this type of energy.

$$\frac{P_1}{\rho g}+\frac{V_1^2}{2g}+Z_1 + H_P = \frac{P_2}{\rho g}+\frac{V_2^2}{2g}+Z_2 \tag{5.18}$$

Where H_p is the head supplied by the device. Similarly,

$$\frac{P_1}{\rho g}+\frac{V_1^2}{2g}+Z_1 - H_T = \frac{P_2}{\rho g}+\frac{V_2^2}{2g}+Z_2 \tag{5.19}$$

Where H_T is the head extracted by the device.

The term velocity head in the energy equation considers the velocity to be uniform over a cross-section of the conduit which is seldom true in real fluids. In any cross-section, velocities of particles will be different and therefore it is necessary to integrate the kinetic energies of all portions of the stream to obtain its total value.

If 'v' is the local velocity through an elementary area da (which is at right angles to v) the mass flow per unit time will be (v da) and the kinetic energy of that mass will be $(\rho \text{ v da}) \text{ v}^2/2$

The total kinetic energy across the entire cross- section

$$\text{A} = \int_A \frac{\rho v^3}{2} \text{ da} = \frac{\rho}{2} \int_A \text{v}^3 \text{da} \qquad (5.20)$$

If the exact velocity profile over a cross section is known, the true kinetic energy can be determined by using Equation (5.20).

It is convenient to use average velocity V and an energy correction factor 'α'; hence the kinetic energy at a section can be written as $\alpha\,(\rho/2)\text{AV}^3$.

Thus the value of the energy correction factor α can be obtained as

$$\frac{\rho}{2} \int_A \text{V}^3 \text{ da} = \alpha \frac{\rho}{2} \text{ AV}^3$$

Solving for α one gets,

$$\alpha = \frac{1}{A} \int_A \left(\frac{v}{V}\right)^3 \text{da} \qquad (5.21)$$

In which v is the local velocity over the area da and V is the average velocity over the cross sectional area A.

The value of 'α' depends on the velocity variation across the area which in turn depends on effects of the viscous shear. Hence 'α' is a measure of effects of the viscous shear on a flow pattern. While applying the energy equation between two sections, one must see whether the velocities at the two sections are uniform or nonuniform.

If the velocity varies over the cross section, the kinetic energy term needs correction and therefore the energy equation takes the form in which α_1 and α_2 are the energy correction factors at sections 1 and 2 respectively.

$$\frac{p_1}{\gamma} + Z_1 + \frac{\alpha_1 v_1^2}{2g} = \frac{p_2}{\gamma} + Z_2 + \frac{\alpha_2 v_2^2}{2g} \tag{5.22}$$

The value of 'α' is greater than unity. The greater the variation in the velocity over a cross section, the greater will be the value of 'α'. For laminar flow in circular pipes, in which the velocity profile is parabolic, the energy correction factor a has a value of 2.0. For turbulent flow, the value of a varies between 1.01 to 1.15, the nomual values being between 1.03 and 1.06. Most fluid flow problems are in the turbulent range of flow for which the value of 'α' is slightly greater than unity. Assuming 'α' to be unity will not, therefore, result in any appreciable error.

For the compressible fluids the Bernoulli's equation takes a different form because of the fact that the mass density does not remain constant but varies with the pressure. The details of this modifications can be referred in Streeter and Wylie.

5.5.3 The Physical Significance of Bernoulli's Equation:

The physical significance of Bernoulli's equations can be easily understood by referring to figure 5.3. In Fig. 5.3 (a), the total head or total energy is a constant for flow without friction and is represented by the total head or total energy line as a horizontal line at a constant distance from the datum plane.

With friction this is not true and the total head line in figure 5.3 (b) is seemed to line to be dropping the right. It will be noticed that when the velocity in the tube increases, the sum of the potential and pressure head must decrease and that for a decreased velocity, the sum of the potential and pressure heads increases.

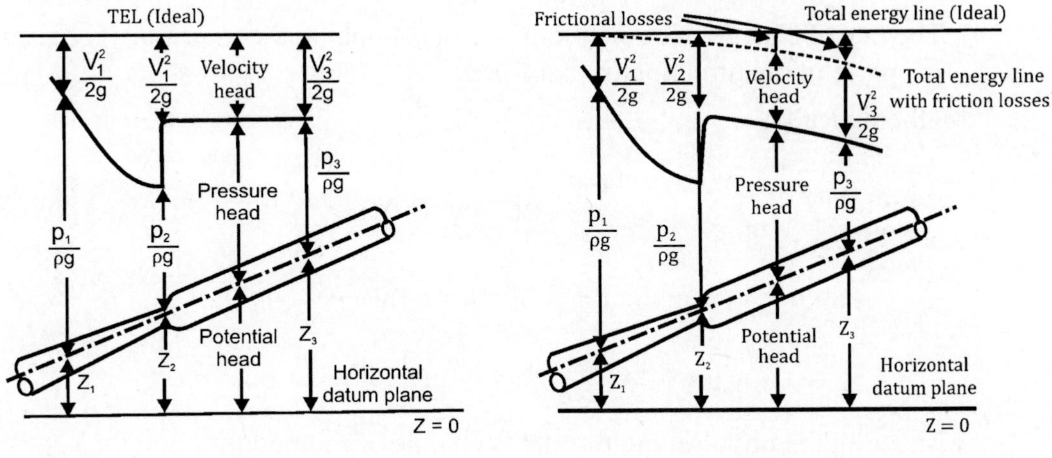

(a) (b)

Fig. 5.3 Head variation in a tube

5.6 Momentum Equation:

The momentum equation is based upon Newton's second law of motion applied to the entire flow system. It provides a simple means of relating the hydrostatic force and the boundary force to the change of momentum flux that is, the momentum which has passed a point during a unit time interval. The force and resulting acceleration are analyzed from the overall viewpoint of the change which occurs, rather than the viewpoint details of the flow processes and internal mechanics. There are number of problems which can be solved with the help of momentum equation such as forces on a pipe bend, the forces of a jet of water impinging on a wall or on the ground and jet propulsion.

5.6.1 Impulse-Momentum Equation:

The momentum of a particle is defined as the product of its mass m and its velocity V.

$$\text{Momentum} = mV$$

The particles of a fluid stream will possess momentum and wherever the velocity of the stream is changed in magnitude or direction, there will be a corresponding change in the momentum of the fluid particles.

In accordance with Newton's second law, a force is required to produce this change which will be proportional to the rate at which the change of momentum occurs.

∴ The rate of change of momentum is proportional to the impressed force and takes place in the direction of that force.

Mathematically,

$$F = \frac{d}{dt}(mV)$$

$$F \cdot dt = d(mV) \tag{5.23}$$

$$F \cdot dt = m \cdot dV$$

where $F \cdot dt$ is impulse and $m \cdot dV$ is change of momentum.

Equation (5.23) can be written as,

$$\text{Force} = \text{mass} \times \text{acceleration}$$

or

$$F = m\left(\frac{V_2 - V_1}{t}\right)$$

$$F = \frac{m}{t}(V_2 - V_1) \tag{5.24}$$

$\frac{m}{t}$ is known as mass flow rate.

$$\frac{\text{mass}}{\text{time}} = \text{Density} \times \frac{\text{Volurne}}{\text{time}}$$

$$\frac{\text{mass}}{\text{time}} = \text{Density} \times \text{Volume rate of flow}$$

$$\frac{\text{mass}}{\text{time}} = \text{Density} \times \text{Discharge} = \rho Q$$

substituting this value in Equation (5.24)

$$F = \rho Q (V_2 - V_1) \tag{5.25}$$

The above equation is termed as momentum equation which shows reaction offered by the surface by virtue of change in momentum.

5.6.2 Momentum Equation for Two Dimensional Flow along a Stream Tube:

It is assumed that fluid is incompressible, the velocity changes uniformly and pipe material is non-elastic.

Fig. 5.4 shows a two dimensional problem in which V_1 makes an angle α_1 with x-axis while V_2 makes an angle a_2.

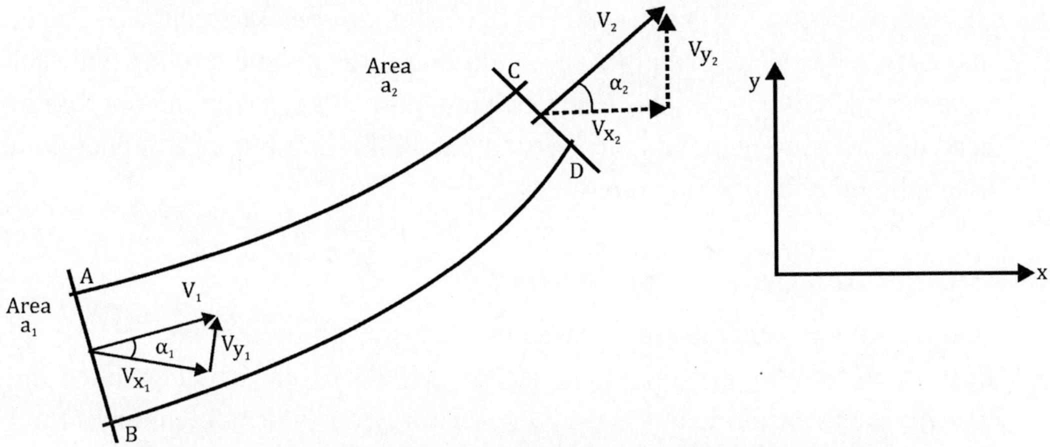

Fig. 5.4 Momentum equation for two dimensional flow

Since both momentum and force are vector quantities, they can be resolved into two components.

Thus F_x and F_y are the components of the resultant force on the element of fluid ABCD.

$$F_x = \text{Rate of change of momentum of fluid in x direction}$$

$$= \rho\, Q\, (V_2 \cos \alpha_2 - V_1 \cos \alpha_1)$$

$$= \rho\, Q\, (V_{x_2} - V_{x_1}) \qquad (5.26)$$

Similarly

$$F_y = \text{Rate of change of momentum of fluid in y direction}$$

$$= \rho\, Q \left(V_{y_2} - V_{y_1}\right) \qquad (5.27)$$

The resultant force is given by,

$$F_R = \sqrt{F_x^2 + F_y^2} \tag{5.28}$$

Direction of resultant force with the horizontal,

$$\tan \theta = \frac{F_y}{F_x} \tag{5.29}$$

The momentum theorem is used especially for solving problems related to forces on bends, elbows etc. in a pipeline, forces on stationary and moving plates or vanes in hydraulic machines, jet propulsion, propellers, in finding out loss of head due transitions in cross sectional areas, in finding out loss of energy in hydraulic jump in open channels.

5.6.3 Momentum Correction Factor β:

Momentum theorem sated above is based on the assumption that the velocity of flow is uniform across the cross section. However, in actual practice the velocity is not uniform: across the cross section. Thus, the momentum of fluid at a section of the passage found out on the basis of average velocity of flow at a section is much different from the actual momentum of a fluid passing through the section. This is due to the variation of velocity across the section of the passage. This necessitated introduction of a correction factor namely the momentum correction factor β. The Concept of momentum correction factor is similar to energy correction factor.

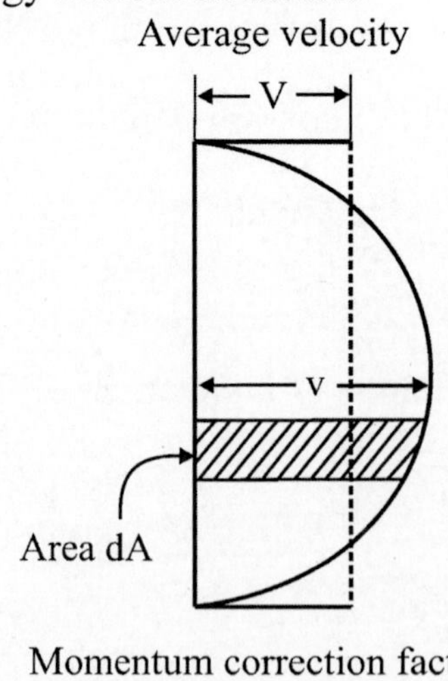

Fig. 5.5 Momentum correction factor (*β*)

Momentum of fluid passing the cross-section per second based on variation of velocity.

$$= \int_A \rho(v \cdot dA) \cdot v = \int_A \rho v^2 dA$$

Momentum of fluid passing the section with area 'A' per second based on average velocity

$$\text{'V'} = \rho\, AV^2$$

Ratio of these two momentums is called as momentum correction factor and is denoted by 'β' (beeta).

$$\therefore \qquad \beta = \frac{\text{[Momentum per second calculated taking into account actual velocity variation]}}{\text{[Momentum per second calculated on the basis of average velocity]}}$$

$$\therefore \qquad \beta = \frac{\int_A \rho v^2 \mathrm{dA}}{\rho A \mathrm{v}^2}$$

$$\beta = \frac{1}{A}\int_A \left(\frac{v}{V}\right)^2 \mathrm{dA} \qquad (5.30)$$

For Laminar flow through circular pipe $\beta = 1.33$. Since for turbulent flow velocity distribution is more or less uniform β for turbulent flow is 1.01 to 1.05. In general, $\alpha > \beta > 1$

5.7 Fluid Flow Measurements:

Bernoulli's equation is one of the important tools for solving many problems in fluid mechanics. It is applied either singly or in combination with the equation of continuity depending upon the results desired.

While applying this equation, the assumptions and limitations discussed earlier should be carefully borne in mind.

Applications of Bernoulli's theorem can be broadly classified in two categories;

(1) Flow through closed conduits

(2) Flows with free surface.

Such a classification of problems does not necessarily change the method of application of Bernoulli's theorem.

There are various devices for measuring the fluid flow which work on the principle of Bernoulli's equation, the discussions on which is beyond the scope of this book. Keeping in view the first course in fluid mechanics Venturimeter, Orifice meter are presented here followed by discussion on pitot tube which is based on Bernoulli's principle and used to measure velocity of flow.

5.8 Venturimeter:

Venturimeter is a device used for measuring the rate of flow of a fluid flowing through a closed conduit. It consists of five parts as shown in figure 5.6

i) **Inlet section:**

It is a starting portion of Venturimeter having same diameter as that of a pipe. There is a pressure ring provided with a pressure tapping for measurement of pressure head at the inlet section.

ii) **Converging cone:**

It is a conical section converging in the direction of flow which reduces the area of flow. The angle between the converging walls of the upstream cone is about 20°.

iii) **Throat:**

It is small tubular portion with uniform cross-section. The diameter of the throat section ranges between $1/3^{rd}$ to $3/4^{th}$ of the diameter of inlet. It is generally $\frac{1}{2}$ of the diameter of inlet. Length of the throut section is equal to its diameter. A pressure ring with pressure tapping is provided at the throat.

iv) **Diverging cone:**

It is a conical tube which diverges gradually in the direction of the flow increasing the area of flow. The angle between diverging walls of this downstream cone is about 6°. The angle of this cone is much smaller than converging cone so that the length of the diverging cone increases. This increased dimension provides sufficient length for the flow to diverge. This avoids eddy formation and in tum the energy losses.

v) Outlet section:

It is the end portion of Venturimeter and has the same diameter as that of the pipe.

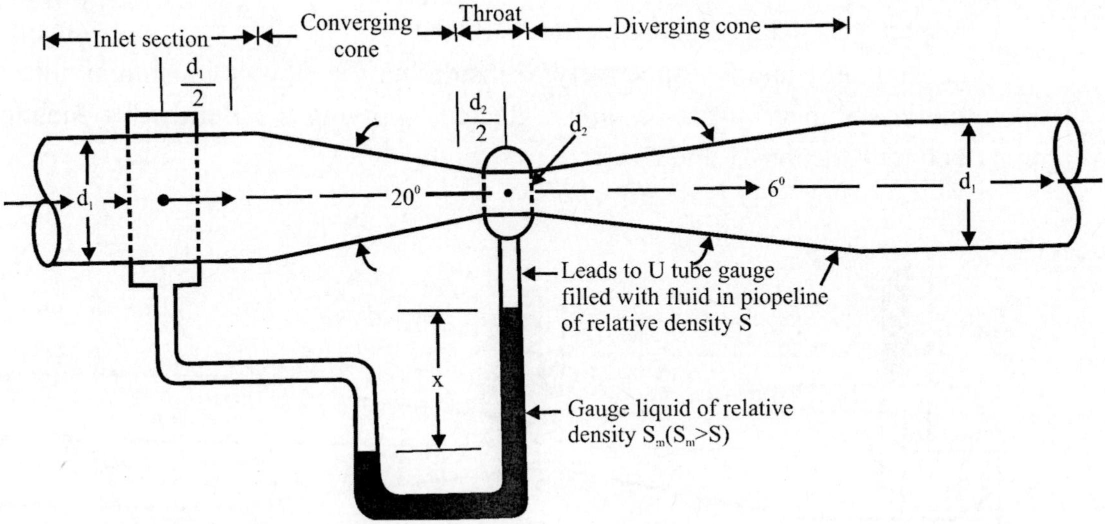

Fig. 5.6 Typical Venturimeter

The size of the venturimeter is expressed in terms of inlet diameter to throat diameter. This ratio called as throat ratio should be such that pressure at throat section does not become negative. To increase the accuracy of discharge measurement the throat diameter is reduced. But this causes velocity to increase at throat, correspondingly a large pressure difference is established between inlet and throat. But on the other hand pressure may become very low at the throat nearing the vapour pressure of the fluid flowing through the pipe. In such cases boiling occurs, vapour bubbles form and they collapse as the fluid moves into a region of higher pressure. This pressure can produce dynamic effects that cause very large pressure transients in the vicinity of bubbles. Pressures as large as 690 Mpa are believed to occur. At this point the liquid gives out dissolved air and begins to vaporize and continuity of flow breaks. This phenomenon is called 'cavitation' and is not desirable. It is always avoided by keeping the throat diameter sufficiently in the range mentioned above.

5.8.1 Expression for Discharge Measurement Through Venturimeter:

Let a_1 and a_2 be the cross-sectional areas at inlet section and throet section respectively, Refer figure 5.7. Let P_1, P_2 and V_1, V_2 be pressures and velocities at sections 1 and 2 respectively. Considering the flow is incompressible, if the losses in venturimeter are neglected, applying the Bernoulli's equation between sections 1 and 2.

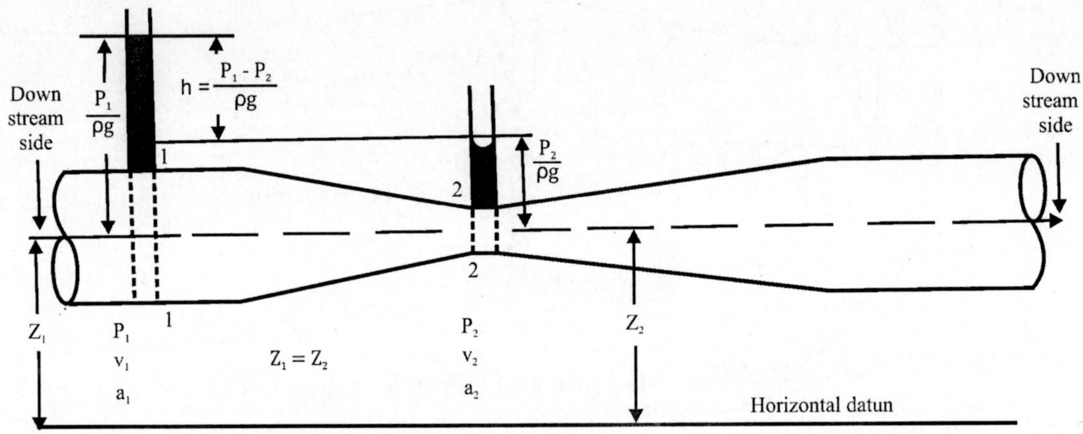

Fig. 5.7 Pressure variation along venturimeter

$$\frac{P_1}{\rho g} + \frac{V_1^2}{2g} + Z_1 = \frac{P_2}{\rho g} + \frac{V_2^2}{2g} + Z_2$$

Since pipe is horizontal $Z_1 = Z_2$

$$\therefore \quad \frac{P_1}{\rho g} + \frac{V_1^2}{2g} = \frac{P_2}{\rho g} + \frac{V_2^2}{2g}$$

or

$$\frac{P_1 - P_2}{\rho g} = \frac{V_2^2 - V_1^2}{2g} = h \tag{5.31}$$

where $\frac{P_1 - P_2}{\rho g} = h =$ Difference in piezometric heads at sections 1 and 2 .

Applying equation of continuity between sections 1 and 2.

$$Q = a_1 V_1 = a_2 V_2$$

or

$$V_1 = \left(\frac{a_2}{a_1}\right) V_2 \tag{5.32}$$

Substituting the value of V_1 in Equation 5.31

$$\frac{V_2^2 - \left(\frac{a_2}{a_1}\right)^2 V_2^2}{2g} = h$$

$$\therefore \quad \frac{V_2^2}{2\,g}\left[1 - \frac{a_2^2}{a_1^2}\right] = h$$

$$\therefore \quad \frac{V_2^2}{2\,g}\left[\frac{a_1^2 - a_2^2}{a_1^2}\right] = h$$

$$\therefore \quad V_2^2 = \frac{a_1^2(2gh)}{a_1^2 - a_2^2} \tag{5.33}$$

$$\therefore \quad V_2 = \frac{a_1\sqrt{2gh}}{\sqrt{a_1^2 - a_1^2}}$$

Since losses are not considered the velocity V_2 in Equation (5.33) gives theoretical value.

∴ Theoretical discharge $\quad Q_{th} = a_2 V_2$

$$\therefore \quad Q_{th} = \frac{a_1 a_2\sqrt{2gh}}{\sqrt{a_1^2 - a_1^2}} \tag{5.34}$$

The Equation 5.34 is often written as,

$$Q_{th} = k\sqrt{h} \tag{5.35}$$

Where

$$k = \frac{a_1 a_2 \sqrt{2g}}{\sqrt{a_1^2 - a_2^2}} \tag{5.36}$$

k is called as Venturimeter constant and is dependent upon the geometry of the Venturimeter. When a real fluid flows through a Venturimeter it is apparent that losses due friction will occur between the inlet and throat section. It will result in a pressure difference between the inlet and throat section which will be somewhat greater than that expected from an ideal fluid and used in the above equation.

As a result, a coefficient is introduced which will relates actual discharge with theoretical discharge. It is called as coefficient of discharge C_d

$$C_d = \frac{\text{Actusl discharge}}{\text{Theoretical discharge}} = \frac{Q_{act}}{Q_{th}} \tag{5.37}$$

Therefore,

$$Q_{act} = C_d \cdot Q_{th} - C_d \frac{a_1 a_2 \sqrt{2gh}}{\sqrt{a_1^2 - a_2^2}} \tag{5.38}$$

It is to be noted that 'h' is the difference of piezometric head of the fluid in the meter. The C_d also accounts for effects of non-uniformity of velocity over sections 1 and 2. Although C_d varies somewhat with flow rate, viscosity of fluid and surface roughness, a value of about 0.98 is usual with fluids of low viscosity.

5.8.2 Vertical *I* Inclined Venturimeter:

A Venturimeter can also be used for measuring the flow rate through a pipe which is held in either vertical or in an inclined position. A Venturimeter connected to an inclined pipe is shown in figure 5.8

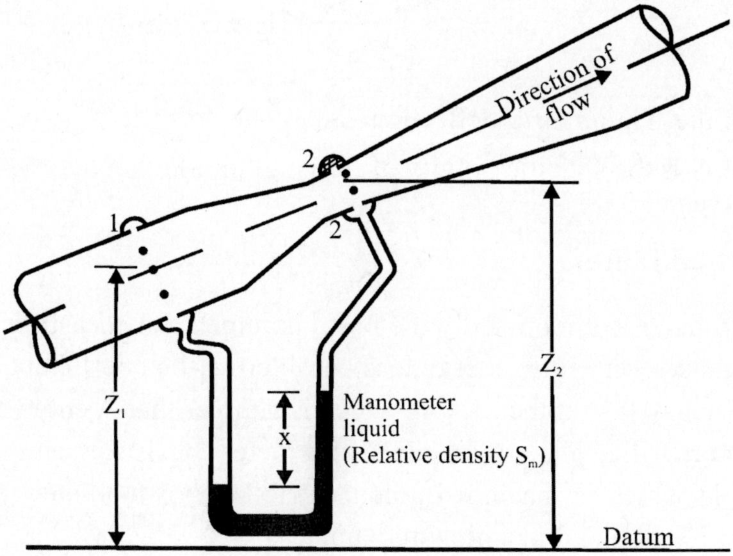

Fig. 5.8 Inclined Venturimeter

Applying Bernoulli's theorem between section 1 and section 2 for no loss of energy.

We get,

$$\frac{P_1}{\rho g} + \frac{V_1^2}{2g} + Z_1 = \frac{P_2}{\rho g} + \frac{v_2^2}{2g} + Z_2$$

$$\therefore \quad \frac{v_2^2 - V_1^2}{2\text{ g}} = \left(\frac{P_1}{Pg} + z_1\right) - \left(\frac{P_2}{Pg} + z_2\right)$$

$$\therefore \quad \frac{V_2^2 - V_1^2}{2\text{ g}} = h \qquad (5.39)$$

The same expression which we obtained in equation (5.30), except the difference of elevation is included in equation (5.31). In figure 5.8 U-tube manometer is used. The value of h is calculated with the help of a fomula

$$h = x\left(\frac{S_m}{S} - 1\right) \text{ Here } S_m > S \qquad (5.40)$$

If inverted U-tube manometer is used

$$h = x\left(1 - \frac{S_m}{S}\right) \text{ Here } S > S_m \tag{5.41}$$

where x is the manometric deflection S_m is the specific gravity of manometric liquid. S is the specific gravity of flowing liquid.

5.9 Orifice Meter:

The Venturimeter is undoubtedly the best instrument for measuring flow in pipes as it causes very little energy loss considering the coefficient of discharge is more than 0.9. Another device which is simpler than Venturimeter is Orifice meter or Orifice plate. A simple orifice meter consists essentially of a circular plate in which is machined a concentric hole of diameter 'd' as shown in figure 5.9 installed in a pipe of diameter D.

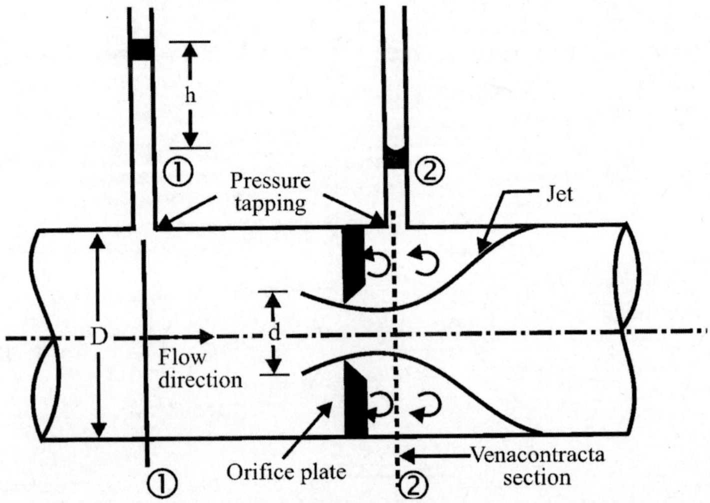

Fig. 5.9 Orificemeter

It is to be observed that the section at which the stream lines come closest together is not at the Orifice plate but at a section approximately at a distance equal to half of the diameter of the plate (d/2) downstream from it. This minimum cross-section of the stream tube is known as vena contracta. Pressure tappings are made at sections 1and 2; wherein section 1 is at a distance of pipe diameter D from the orifice and section 2 is at vena contracta. The pressure at point 2 is minimum and velocity maximum, as the area of vena contracta is minimum.

Downstream of section 2 the flow lines break down into a highly turbulent area wherein the flow regains its original diameter. The area of vena contracta a_c is less than area of orifice a.

Applying Bernoulli's theorem to sections 1 – 1 and 2 – 2, we get,

$$\frac{P_1}{\rho g}+\frac{V_1^2}{2g}+Z_1=\frac{P_2}{\rho g}+\frac{v_2^2}{2g}+Z_2 \quad [\text{where } Z_1 = Z_2]$$

$$\therefore \qquad \frac{V_2^2-V_1^2}{2\,g}-\frac{P_1-P_2}{\rho g}=h \qquad (5.42)$$

where h is the pressure head difference between sections 1 – 1 and 2 – 2 in terms of liquid flowing through the pipe and can be determined using a piezometers or U-tube manometer. By continuity equation we can write,

$$aV_1 = a_C V_2 \qquad (5.43)$$

where A is area of pipe and a_C is area of jet at vena contracta.

$$\frac{a_c}{a} = C_c \qquad (5.44)$$

where a is area of orifice and C_C is called as coefficient of contraction for the orifice.

$$a_c = a \cdot C_c$$

$$\therefore \qquad aV_1 = a \cdot C_c \cdot V_2 \qquad (5.45)$$

$$\therefore \qquad V_1 = C_c \cdot \frac{a}{A} \cdot V_2$$

Substituting value of V_1 in Equation 5.34

$$v_2^2 - C_c^2\left(\frac{a}{A}\right)^2 v_2^2 = 2gh$$

$$V_2 = \sqrt{\frac{2gh}{1-C_c^2\left(\frac{a}{A}\right)^2}} \tag{5.46}$$

This is the theoretical velocity at vena contracta since losses have not been taken into account. To account for losses, coefficient of velocity C_V is introduced.

$$C_c = \frac{\text{actual velocity}}{\text{theoretical velocity}} \tag{5.47}$$

$$\therefore \quad V_{2\ \text{actual}} = C_v \cdot V_{2}\ \text{theoretical}$$

$$V_{2\ \text{actual}} = C_v \sqrt{\frac{2gh}{1-c_c^2\left(\frac{a}{A}\right)^2}} \tag{5.48}$$

Actual discharge $\quad Q = C_e \cdot a \cdot V_2$

$$\therefore \quad Q = \frac{C_c C_v}{\sqrt{1-C_c^2\left(\frac{a}{A}\right)^2}} \cdot a \cdot \sqrt{2gh} \tag{5.49}$$

But $\quad C_c \cdot C_V = Cd \qquad (5.50)$

$$\therefore \quad Q = \frac{Cd}{\sqrt{1-C_c^2\left(\frac{a}{A}\right)^2}} \cdot a \cdot \sqrt{2gh} \tag{5.51}$$

Defining, coefficient of orifice meter

$$C = \frac{1}{\sqrt{1-C_c^2\left(\frac{a}{A}\right)^2}} \tag{5.52}$$

$\therefore$ The equation can be written as,

$$Q = Cd \cdot C \cdot a\sqrt{2gh} \tag{5.53}$$

In general, for high values of discharge the coefficient of discharge is approximately 0.62 to 0.65 indicating that losses are fairly high. This endorses the earlier stamen that Venturimeter is a better equipment as far as accuracy is concerned in that it has coefficient of discharge greater than 0.9 meaning losses are less.

5.10 Pitot Tube:

The velocity at a point or a number of points throughout a section in a fluid stream is often needed in order to establish the velocity profile. This velocity profile may be used to obtain the average velocity throughout the section from an integration of the velocity profile in order to determine the flow rate in the fundamental studies of boundary layers or wakes. A point velocity is almost impossible to measure since any sensing device occupies a finite region. However if the area of the flow occupied by the sensing device is very small compared with the total area of the flow stream the measured velocity can be considered to be a point velocity.

A useful concept associated with Bernoulli's equation deals with the dynamic and stagnation pressures.

The Bernoulli's equation cam also be written in the form,

$$P + \frac{1}{2}\rho V^2 + \rho gz = \text{constant} \tag{5.54}$$

The first term, *P* is the actual thermodynamic pressure of the fluid as it flows. To mensure its value one can move along with the fluid thus being static relative the moving fluid. Henoe it is normally termed as static pressure. The third term in the equation (5.46) ρgz is termed as hydrostatic pressure. It represents a pressure due to potential energy, by virtue of its elevation. The second term in the equation (5.46) $\frac{1}{2}\rho V^2$ is termed as dynamic pressure.

Refer figure 5.10. The point 2 is called as stagnation point as velocity of fluid at that point is zero.

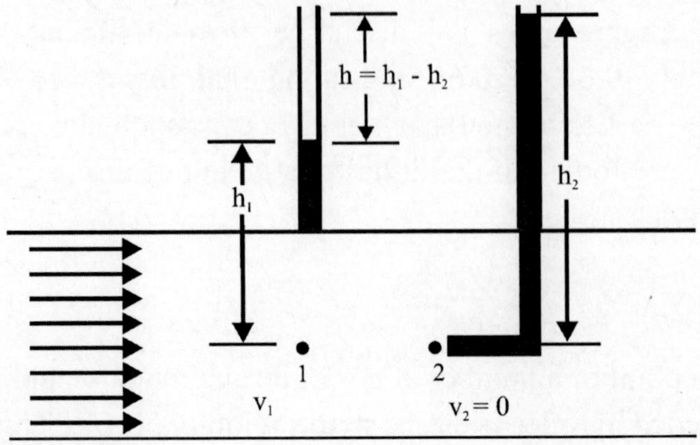

Fig. 5.10 Measurement of static and stagnation pressure

Applying Bernoulli's equation between points (1) and (2).

$$P_1 + \frac{\rho V_1^2}{2} = P_2 + \frac{\rho V_2^2}{2} \quad (Z_1 = Z_2)$$

At point 2, $V_2 = 0$

$$\therefore \quad P_2 = P_1 + \frac{\rho v_1^2}{2} \tag{5.55}$$

Hence pressure at stagnation point (2) is greater than the static pressure at point (1) by an amount $\frac{\rho V_1^2}{2}$, the dynamic pressure which is also evident from Bernoulli's equation.

The Equation (5.47) can be written as

Stagnation pressure = Static pressure + Dynamic pressure

If the elevation effects are neglected, the stagnation pressure is the largest pressure obtainable along a stream line. It represents the conversion of all the kinetic energy into a pressure rise. This principle is used in construction of pitot tube (Fig. 5.11). Pitot tube provides one of the most accurate means of measuring the velocity of fluid flow. For an open stream of liquid only single tube is necessary as shown in figure 5.11. Applying Bernoulli's equation we get,

$$V_1 = \sqrt{2gh} \qquad (5.56)$$

If the stream velocity at a point in a pipe is to be measured the same equation is obtained

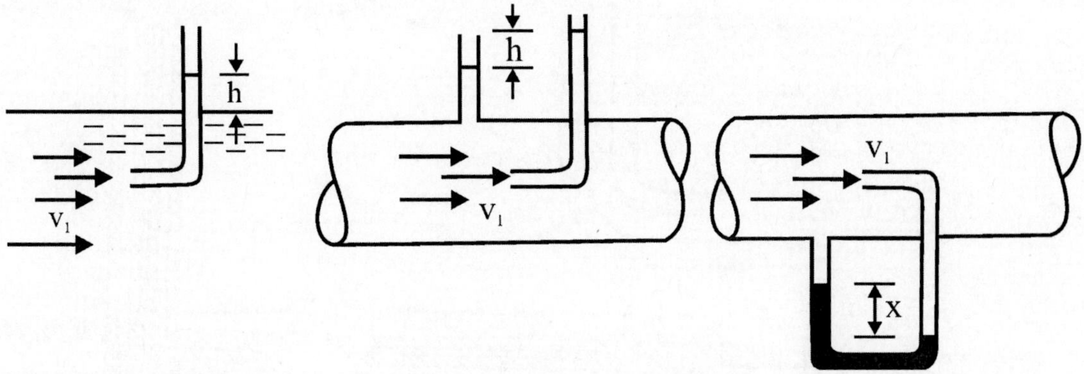

(a) in open channel **(b) in a pipe** **(c) in a pipe connected to a manometer**

Fig. 5.11 Pitot tube

In practice, it is difficult to measure the height 'h' above the surface of the moving liquid. The tubes in 5.11 (b) may be connected to a differential U-tube manometer as shown in figure 5.11(c) to result in a more convenient system. The 'h' is obtained very easily as in case of Venturimeter.

$$h = x\left(\frac{S_m}{S} - 1\right)$$

$$\therefore \quad V = \sqrt{2gx\left(\frac{s}{s} - 1\right)} \qquad (5.57)$$

5.10.1 Pitot Static Tube:

Sometimes, both static pressure measuring tube and the stagnation pressure measuring tube are combined in one device called pitot static tube as shown in figure 5.12 (a) and (b).

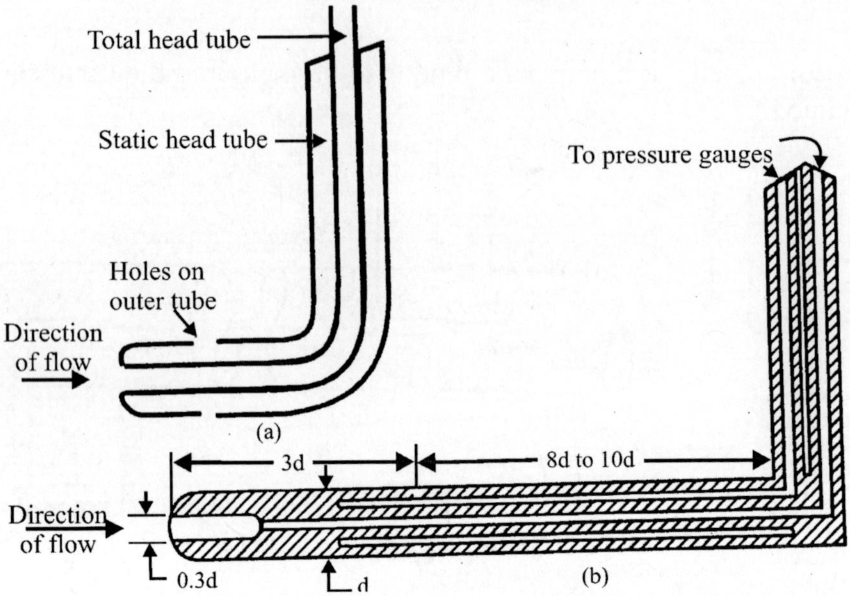

(a) Pitot static tube **(b) Prandtl's-Pitot tube**

Fig. 5.12 Pitot static tube

It consists of a small, cylindrical tube surrounded by a closed outer tube with annular space in between them. The opening at the inner tube records stagnation pressure while the outer tube with holes drilled on it measures the static pressure. The pressure head difference can be measured by connecting the outlets from the pitot-static tube to the ends of a U-tube differential manometer. Figure 5.12 (b) shows some specific dimensions as suggested by Prandtl, which is known as Prandtl's-Pitot tube.

The foregoing sections presented determination of position, velocity and acceleration of fluid particle at a particular time 't' using the force responsible for it by using the Bernoulli's equation. The modified form of Bernoulli's equation was also presented along with its applications especially for discharge and velocity measuring equipment. The momentum equation was also derived which is another way of solving the fluid flow problem.

5.11 Solved Examples:

Ex. 5.1 : A reducing bend is placed in A pipeline such that the direction of flow of water is turned through 60° in the horizontal plane and the pipe diameter is reduced from 0.25 m to 0.15 m. The velocity and pressure at the entry to the bend are 1.5 m/s and 300 kN/m^2 gauge respectively. At the exit the pressure is 287.2 KN/m^2 gauge.

i) Determine the force exerted by the bend on the water.

ii) What would be the force of the water on the bend?

Solution:

Angle of bead $= \theta = 60°$

Bend is in horizontal plane i.e. $Z_1 = Z_2$

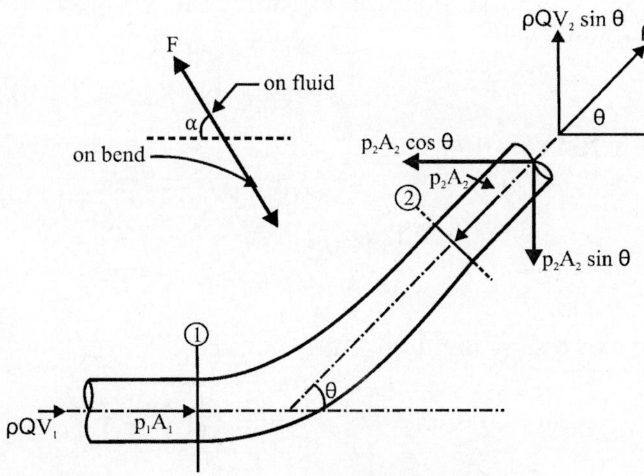

Fig. Ex. 5.1

$$A_1V_1 = A_2V_2$$

$$\frac{\pi d_1^2}{4} \times V_1 = \frac{\pi d_2^2}{4} \times V_2$$

$$\frac{\pi \times 0.25^2}{4} \times V_1 = \frac{\pi \times 0.15^2}{4} \times V_2$$

$$0.0491\ V_1 = 0.0177\ V_2$$

	$\therefore \quad V_1 = 0.3599\ V_2$ But $V_1 = 1.5$ m/sec $\therefore \quad V_2 = 4.17$ m/sec $\therefore \quad Q = A_1V_1 - 0.0491 \times 1.5$ $= 0.07365\ m^3/sec$ The momentum equation in x-direction is given by equation $P_1 A_1 - P_2 A_2 \cos\theta + F_x = Q (V_2 \cos\theta - V_1)$ $300 \times 10^3 \times 0.0491 - 287.2 \times 10^3 \times 0.0177 \cos 60^0 + F_x$ $= 1000 \times 0.07365 (4.17 \cos 60^0 - 1.5)$ $\therefore \quad F_x = -12145.19$ N Similarly, the momentum equation in y-direction is given by equation $-P_2 A_2 \cos\theta + F_y = \rho Q (V_2 \sin\theta - 0)$ $287.2 \times 10^3 \times 0.0177 \cos 60^0 + F_y$ $= 1000 \times 0.07365 (4.17 \sin 60^0 - 0)$ $\therefore \quad 2541.72 + Fy = 265.974$ $\therefore \quad Fy = -2275.745$ Force exerted by bend on fluid $= F = \sqrt{F_x^2 + F_y^2}$ $= \sqrt{(-12145.19)^2 + (-2275.745)^2}$ $= 12356.56$ N An equivalent force would be acted on by water on bend.

Ex. 5.2 :	A 90^0 bend in a 15 cm diameter pipe carries oil of specific gravity 0.8 at 110 lit/sec under a pressure of 0.8 m of oil at the entrance. Find the force on the bend.

Solution:

Angle of bend = angle through which water get deflected = $\theta = 90^0$

$d_1 = d_2 = 15$ cm = 0.15 m

$S_0 = 0.8$

Q = 110 lit/sec = 110 x 10^{-3} m^3/sec = 0.110 m^3/sec

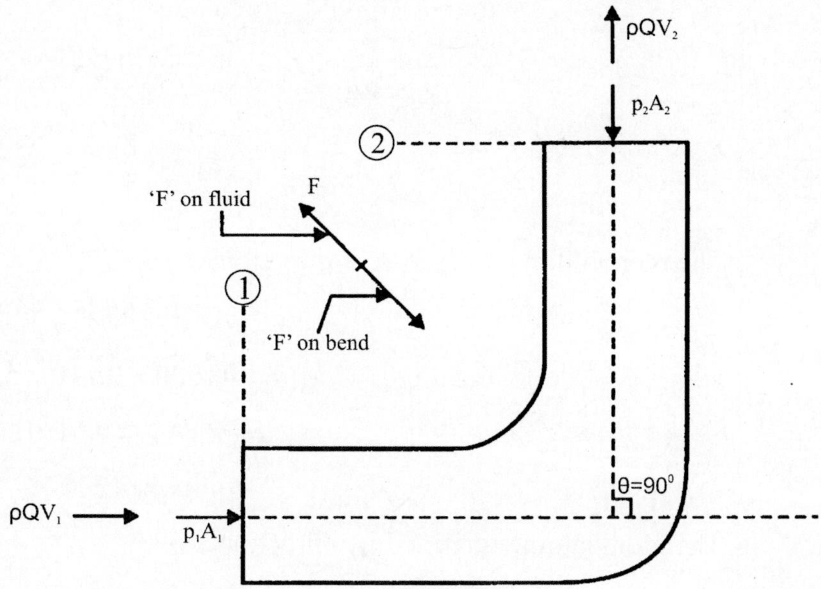

$P_1 = 0.8$ m of oil

Fig. Ex. 5.2

Assuming the bend is in a horizontal plane i.e. $Z_1 = Z_2$
Using continuity equation

$$Q = A_1V_1 - A_2V_2$$

∴ $$A_1 = A_2 \longrightarrow V_1 = V_2$$

∴ $$Q = \frac{\pi d_1^2}{4} \times V_1$$

$$0.110 = \frac{\pi \times 0.15^2}{4} \text{ x } V_1$$

$\therefore$ $$0.110 = 0.0177\ V_1$$

$\therefore$ $$V_1 = 6.22 \text{ m/sec}$$

Applying Bernoulli's equation at 1 and 2 :

$$\frac{P_1}{\gamma} + \frac{v_1^2}{2g} + Z_1 = \frac{P_2}{\gamma} + \frac{V_2^2}{2g} + Z_2$$

$\therefore$ $$V_1 = V_2\ Z_1 - Z_2$$

$$P_1 = P_2$$

The momentum equation in x-direction :

$$P_1\ A_1 + F_x = \rho Q\ (0 - V_1)$$

$$0.8 \text{ x } 0.0177 + F_x = -1000 \text{ x } 0.110 \text{ x } 6.22$$

$\therefore$ $$F_x = -684.2 - 0.01416$$

$$= -684.2142 \text{ N}$$

The momentum equation in y-direction

$$-P_2\ A_2 + F_y = \rho Q\ (V_2 - 0)$$

$$-0.8 \text{ x } 0.0177 + F_y = +1000 \text{ x } 0.110 \text{ x } 6.22$$

$\therefore$ $$F_y = 684.2142 \text{ N}$$

$\therefore$ $$\text{Force on fluid} = \sqrt{F_x^2 + F_y^2}$$

$$= \sqrt{(684.2142)^2 + (+6842142)^2}$$

$$= 967.62 \text{ N}$$

$\therefore$ Force exerted on the bend will have the same magnitude but opposite direction to F

Ex. 5.3 : The centre line of tapered pipe AB slopes down from A to B at an angle of 30^0 to the horizontal. The distance AB is 5 m and the diameter increases uniformly from 100 mm at A to 150 mm at B. The pipe carries petrol (S = 0.74) and pressure gauges are installed at A and B. Find (a) the flow rate when the reading on the pressure gauges are equal, (b) the pressure difference across AB for the same flow rale when the direction of taper is reversed.

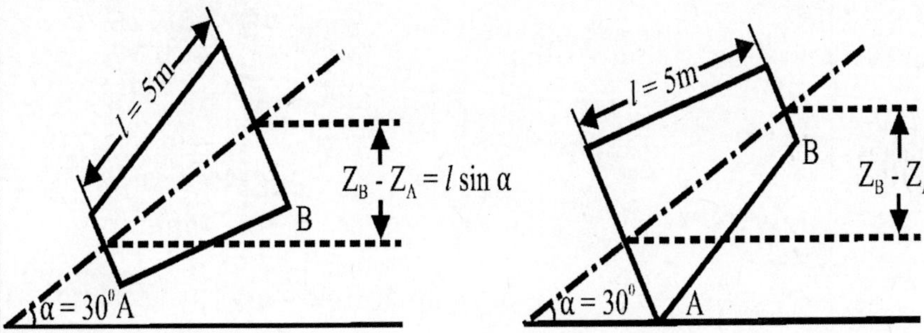

Fig. Ex. 5.3

Solution:

Angle of inclination $\alpha - 30°$

l(AB) = 5 m

Diameter at A = 100 mm

Diameter at B = 150 mm

Specific gravity (S) = 0.74

$Q = V_A a_A = V_B a_B$ Applying the Bernoulli's equation between A and B.

$$\frac{P_A}{S\rho g} + \frac{v_A^2}{2g} + Z_A = \frac{P_B}{S\rho g} + \frac{v_B^2}{2g} + Z_B$$

$$\left[\text{Here } \frac{P_A}{S\rho g} = \frac{P_B}{S\rho g} = \text{ Given}\right]$$

$$Z_B - Z_A - \frac{v_A^2 - v_B^2}{2g} = \frac{v_B^2}{2g}\left[\frac{v_A^2}{v_B^2} - 1\right]$$

$$l \sin \alpha = \frac{v_B^2}{2g}\left[\frac{a_A^2}{a_B^2} - 1\right]$$

$$5 \sin 30^0 = \frac{v_B^2}{2\,g}\left[\left(\frac{0.15}{0.1}\right)^4 - 1\right]$$

$$2.5 = \frac{v_B^2}{2\times 9.81}[5.0625 - 1]$$

$$2.5 = \frac{v_B^2 \times 4.0625}{2\times 9.81}$$

$$v_B^2 = \frac{2.5}{0.207} = 12.0738$$

$$V_B = 3.474 \text{ m/s}$$

$$Q = a_B v_B = \frac{\pi}{4}(0.15)^2 \times 3.474 = 0.0614 \text{ m}^3/\text{s}$$

$$= 61.4 \text{ liters/s}$$

$$Q_A V_A = a_B v_B \times 3.747 = 0.0614 \text{ m3/s}$$

$$0.0614 = (0.1)2 \times v_B$$

∴ $v_B = 7.82$ m/s

and $0.0614 = \frac{\pi}{4}(0.15)^2 \times v_A$

∴ $v_A = 3.474$ m/s

Applying the Bernoulli's equation between points A and B.

$$\frac{p_A}{Spg} + \frac{v_A^2}{2\,g} + Z_A = \frac{p_B}{S\rho g} + \frac{v_B^2}{2\,g} + Z_B$$

$$\frac{p_B - p_A}{S\rho g} = \frac{v_A^2 + v_B^2}{2\,g} + (Z_A - Z_B)$$

$$= 0.615 - 3.11684 - 2.5 = -2.5 - 2.5 = -5$$

or $P_A - P_B = 0.74 \times 1000 \times 9.81 \times 5$

$$= 36297 \text{ N/m}^2 = 36.297 \text{ kPa}$$

Ex. 5.4:	A. Francis turbine has a vertical draft tube. The diameter of the tube on the upper side is connected to the outlet of the turbine runner and through which water enters is 600 mm and that of the outlet is 900 mm. The tube is running full with water flowing downwards and it is 6 m long with 1 m of its bottom length drowned in tail race. The frictional loss in the vertical draft tube is K $\left(\frac{v_1^2-v_2^2}{2g}\right)$ where K is efficiency of conversion 90%. The velocity at the entrance to the draft tube is 8 m/s. Find the pressure at the entrance of the draft tube. Take datum at the water surface. 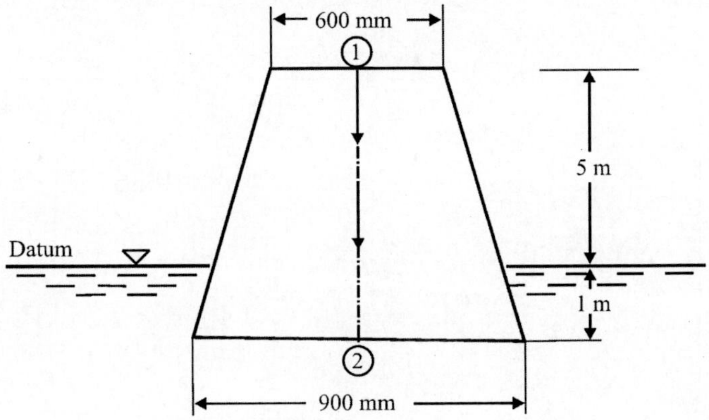 **Fig. Ex. 5.4** **Solution:** Diameter $D_1 = 600$ mm (inlet) $D_2 = 900$ mm (outlet) Length of tube = 6 m + 1 m in drowned condition. Frictional loss = $K\left(\frac{v_1^2-v_2^2}{2\,g}\right)$ $V_1 = 8$ m/s $a_1V_1 = a_2V_2$ $\frac{\pi}{4}(0.6)^2 \times 6 = \frac{\pi}{4}(0.9)^2 \times V_2$

	$V_2 = \dfrac{1.696}{0.636} = 2.66$ m/s $h_f = k\left(\dfrac{v_1^2 - v_2^2}{2g}\right)$ $0.9\left[\dfrac{(6)^2-(2.66)^2}{2\times9.81}\right] = 0.9 \times 1.474 = 1.3268$ Putting $Z_1 = 5$ m, $Z_2 = -1$ m (below the datum line). Applying Bernoulli's equation between point 1 and 2. $\dfrac{P_1}{\rho g} + \dfrac{v_1^2}{2\,g} + Z_1 = \dfrac{P_2}{\rho g} + \dfrac{v_2^2}{2g} + Z_2 + h_1$ $\dfrac{P_1}{\rho g} + \dfrac{(6)^2}{2\times9.81} + 5 = 0 + \dfrac{(2.66)^2}{2\times9.81} - 1 + 1.3268$ $\dfrac{P_1}{\rho g} = 0.36 - 1 - 5 - 1.83486 + 13268$ $= -6.148$ m of water $P_1 = -60.25$ kPa $= 60.25$ kPa (vacuum)

Ex. 5.5:	In a vertical pipe conveying kerosene (s = 0.8), pressure gauges are inserted at A and B, where the diameters are 150 mm and 75 mm respectively. The point B is 3 m below A and when the rate of flow down the pipe is 20 liters/s. The pressure at B is 9 kPa greater than at A. Assuming that the losses in the pipe between A and B can be expressed as $Kv_\rho^2/2g$ where V_A is the velocity at A, find the value of k.

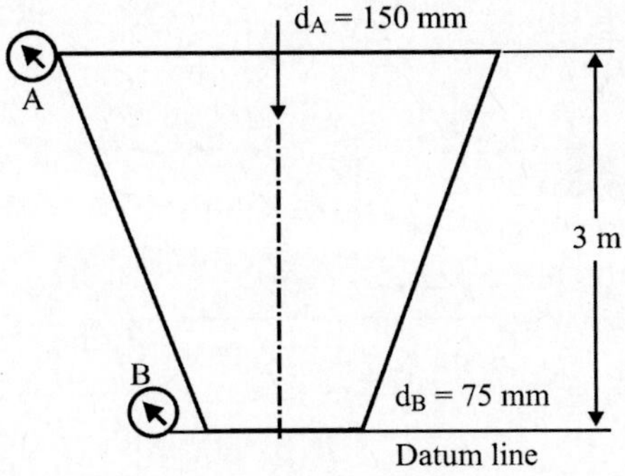

Fig. Ex. 5.5

Solution:

Specific gravity of kerosene s = 0.8

$D_A = 150$ mm, $D_B = 75$ mm

$l(AB) = 3$ m, B is below A

Q = 20 lit/sec

Pressure at B = Pressure at A + 9 kPa

$p_B - p_A = 9$ kPa

$$\frac{p_B - p_A}{S\rho g} = \frac{9}{0.8 \times 9.81}$$

= 1.146 m of kerosene

From continuity equation:

$$Q = a_A V_A = a_B V_B$$

$$V_A = Q/a_A = \frac{0.02}{\pi/4(0.15)^2} = 1.13 \text{ m/s}$$

$$V_B = Q/a_B = \frac{0.02}{\pi/4(0.15)^2} = 4.527 \text{ m/s}$$

	Applying the Bernoulli's equation between points A and B: $$\frac{P_A}{S\rho g}+\frac{v_A^2}{2g}+Z_A=\frac{P_B}{S\rho g}+\frac{v_B^2}{2g}+Z_A+Z_B+kv_A^2/2g$$ $$\frac{P_A-P_B}{S\rho g}+\frac{v_A^2-v_B^2}{2\,g}+(Z_A-Z_B)=k(v_A^2/2g)$$ $$-1.146+0.065-1.044+3=k(0.065)$$ $$3.065-2.19=k(0.065)$$ $$0.875=k(0.065)$$ $\therefore$ $$k=13.46$$

Ex. 5.6 :	Water is pumped at the rate of 300 litres/sec through a 30 cm pipe upto a hill wop. On the hill top which has an elevation of 50 m, the diameter of pipeline reduces to 20 cm. If the pump maintains a pressure of 981 bar at the hill top, what is the pressure at the foot hills having zero elevation? What is the power required to pump the water? **Solution:** Discharge Q = 300 lit/sec = 0.3 m^3/s Diameter D_1 = 30 cm = 0.3 m Elevation Z_2 = 50 m Diameter D_2 = 20 cm = 0.2 m Pressure P_2 = 981 bar = 980 x 10^2 N/m^2 Discharge Q = 300/1000 = 0.3 $m^3/5$ Velocity of flow at 30 cm section, $$V_1=\frac{0.3}{\frac{\pi}{4}(0.3)^2}=4.24\text{ m/s}$$

	Velocity of flow at 20 cm section, $$V_2 = \frac{0.3 \times 4}{\pi(0.2)^2} = 9.55 \text{ m/s}$$ Applying. Bernoulli's equation between the pipeline sections at the foot hills and at the hill top, neglecting energy losses in between, noting that $$1 \text{ bar} = 100 \text{ N/m}^2,$$ $$0 + \frac{p_1}{\gamma} + \frac{(4.24)^2}{2 \times 9.81} = 50 + \frac{981 \times 100}{9810} + \frac{(9.55)^2}{2 \times 9.81}$$ $$= 64.65 \text{ m of water}$$ resulting in, $$P_1 = 63.734 \times 9810 \text{ N/m}^2$$ $$= 625.2 \text{ kN/m}^2$$ $$= 6252 \text{ bar}$$ Pressure at the foot hills = 6252 bar Power required to pump water to the hill top. $$= \gamma QH$$ $$= 9810 \times 0.3 \times 64.65$$ $$= 190265 \text{ Nm/s}$$ $$= 190265 \text{ W}$$ $$= 190.265 \text{ kW}$$

Ex. 5.7:	A pipe 20 cm in diameter is connected to a water tank as shown in Fig Ex 5.7. A nozzle fitted at the end of the pipe discharges into the atmosphere. Calculate the flow rate and the pressure at A, B, C and D. Neglect losses. The diameter of the nozzle throat is 5 cm.

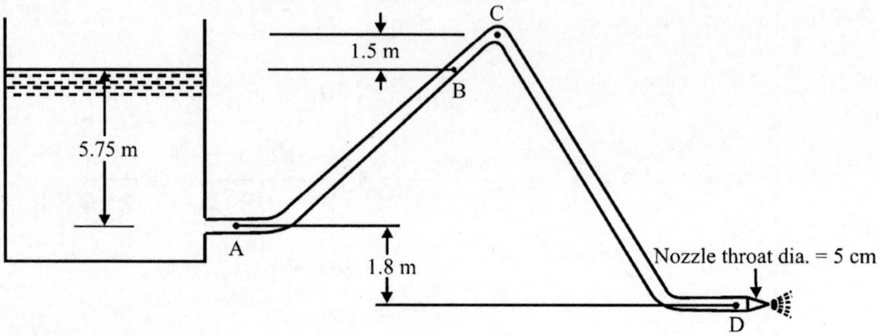

Fig. Ex. 5.7

Solution:

Diameter of pipe 20 cm

Diameter of nozzle throat 5 cm

Figures shows difference of elevation between AB = 5.75 m, BC = 1.5 m

Applying Bernoulli's equation between the water surface in the tank and the jet of water issued from the nozzle, taking the nozzle axis as the datum, and neglecting losses,

$$(5.75 + 1.8) + 0 + 0 = \frac{V^2}{2\,g} + 0 + 0$$

$\therefore$ Velocity of jet, $\quad V = \sqrt{2 \times 9.81 \times 7.55} = 12.17$ m/s

The flow rate, $\quad Q = \frac{\pi}{4}\left(\frac{5}{100}\right)^2$ x 12.17 = 23.9 liters/sec

Velocity in the 20 cm pipe

$$= \frac{0.0239 \times 4}{\pi \times (0.2)^2} = 0.76 \text{ m/s}$$

The pressures at A, B, C and D may be obtained by applying Bernoulli's equation between the water surface in the tank and these points.

i) Pressure at A : $5.75 + 0 + 0 = \frac{(0.76)^2}{2\times9.81} + \frac{P_A}{\gamma}$

$\therefore \quad \frac{P_A}{\gamma} = 5.7205$ m of water

$\therefore \quad P_A = 5.72 \times 9810 \text{ N/m}^2 = 56.1 \text{ kN/m}^2$

$= 561$ bar

ii) Pressure at B : $0 + 0 + 0 = 0 + \frac{(0.76)^2}{2\times9.81} + \frac{P_B}{\gamma}$

$\therefore \quad \frac{P_B}{\gamma} = -0.0295$ m of water

$\therefore \quad P_B = -0.0295 \times 9810$

$= -289.4 \text{ N/m}^2$

iii) Pressure at C : $0 + 0 + 0 = 1.5 + \frac{(0.76)^2}{2\times9.81} + \frac{P_c}{\gamma}$

$\therefore \quad \frac{P_c}{\gamma} = -1.529$ m of water

$\therefore \quad P_C = -1.5295 \times 9810 = -15.0 \text{ kN/m}^2$

iv) Pressure at : $5.75 + 1.8 + 0 + 0 = \frac{(0.76)^2}{2\times9.81} + \frac{P_d}{\gamma} + 0$

$\frac{P_d}{\gamma} = 7.5205$ m of water

$= 7.5205 \times 9810 \text{ N/m}^2$

$= 73.8 \text{ kN/m}^2$

Ex. 5.8 : Water moves steadily through the turbine shown in the Fig. Ex. 5.8 at the rate of 0.23 m^3/s. The pressures at (1) and (2) are 186.4 kN/m^2 and –19.6 kN/m^2 respectively. Neglecting heat transfer, determine the horsepower delivered to the turbine from water.

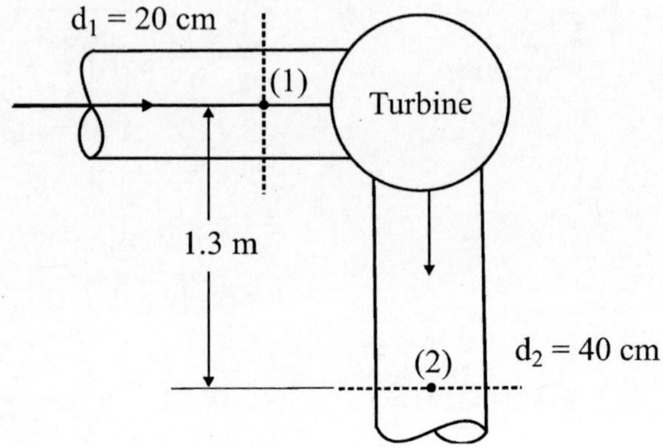

Fig. No. 5.8

Solution:

Discharge $$Q = \frac{0.23\ m^3}{s}$$

Pressures at (1) and (2) = 186.4 kN/m^2 and –19.6 kN/m^2

Velocity at the 20 cm section:

$$-\frac{0.23\times4}{\pi(0.2)^2} = 7.32\ m/s$$

Velocity at the 40 cm section

$$= \frac{0.23\times4}{\pi\times(0.4)^2} = 1.83\ m/s$$

Energy head available at (1) with reference to horizontal datum passing section (2):

$$= 1.30 + \frac{186.4\times10^3}{9810} + \frac{(7.32)^2}{2\times9.81} = 23.035\ Nm/N$$

Energy head in the flow at section (2) :

$$= 0 - \frac{19.6\times10^3}{9810} + \frac{(1.83)^2}{2\times9.81} = -2.0 + 0.171$$

$$= -1.829\ Nm/N$$

	Energy head utilised by the turbine, $= 23.025 - (-1.829)$ $= 24.864$ Nm/N Power delivered to turbine: $= \gamma Q \times 24.864 = 9810 \times 0.23 \times 24.864$ $= 56 \times 10^3$ W $= 56$ kW

Ex. 5.9:	A 120^0 bend cum reducer has 300 mm diameter at inlet and 200 mm diameter at the outer end. When it carries a flow 0.3 m^3/s of water, the pressure at that inlet section is 210 kN/m^2. Assuming no energy loss in the bend determine the force exerted by the water on the bend. The bend is in a horizontal plane. 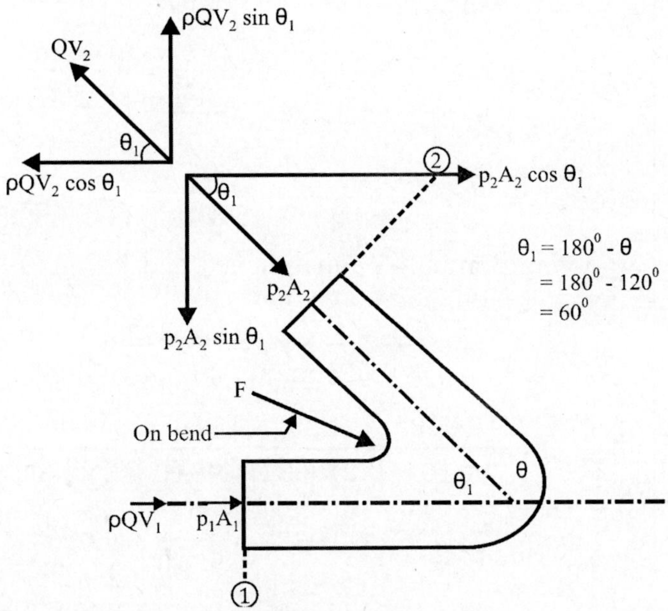 **Fig. Ex. 5.9** **Solution:** Angle of bend $= \theta = 120^0$ $d_1 = 300$ mm $= 0.3$ m $d_2 = 200$ mm $= 0.2$ m $Q = 0.3$ m^3/sec

$P_1 = 210\ kN/m^2 = 210 \times 10^3\ N/m^2$

Bend is in a horizontal plane i.e. $Z_1 = Z_2$

Using continuity equation:

$$Q = A_1 V_1 = A_2 V_2$$

$$\therefore \quad V_1 = \frac{Q}{A_1}$$

$$= \frac{Q}{\frac{\pi d_1^2}{4}} = \frac{0.3}{\left(\frac{\pi \times 0.3^2}{4}\right)}$$

$$\text{and} \quad V_2 = \frac{\frac{Q}{A_2}}{0.07} = 4.29\ m/sec$$

$$= \frac{Q}{\frac{\pi d_2^2}{4}} = \frac{0.3}{\left(\frac{\pi \times 0.2^2}{4}\right)}$$

$$= \frac{0.3}{0.03} = 10\ m/sec$$

Applying Bernoulli's equation at 1 and 2 :

$$\frac{P_1}{\gamma} + \frac{V_1^2}{2g} + Z_1 = \frac{P_2}{\gamma} + \frac{v_2^2}{2g} + Z_2$$

$$\frac{210 \times 10^3}{9810} + \frac{4.29^2}{2 \times 9.81} = \frac{P_2}{9810} + \frac{10^2}{2 \times 9.81}$$

$$21.4067 + 0.9380 = \frac{P_2}{9810} + 5.0968$$

$$\therefore \quad P_2 = 169201.899\ N/m^2$$

The momentum equation in x-direction :

$$P_1 A_1 + P_2 A_2 \cos\theta_1 + F_X = \rho Q(-V_2 \cos\theta_2 - V_1)$$

$$210 \times 10^3 \times 0.07 + 169201.899 \times 0.03 \cos 60^0 + F_X$$

$$= 1000 \times 0.3(-10 \cos 60^0 - 4.29)$$

$$\therefore \quad 14700 + 2538.03 + F_X = -2787$$

	∴ $F_X = -20025.03$ N The momentum equation in y-direction $$P_2 A_2 \sin\theta_1 + F_y = \rho Q (V_2 \sin\theta_1 - 0)$$ $$-169201.899 \times 0.03 \times \sin 60^0 + F_y = 1000 \times 0.3\ (10 \sin 60^0)$$ $$-4395.99 + F_y = 6994.07 \text{ N}$$ ∴ Force acting on the fluid $= F = \sqrt{F_x^2 + F_y^2}$ $$= \sqrt{(-20025.03)^2 + (6994.07)^2}$$ $$= 21211.29 \text{ N}$$ The force exerted by the water on the bend will have same magnitude but opposite direction to ' F '.

Ex.5.10: A pipe of varying section has a sectional area of 3000, 6000 and 1250 mm^2 at point A, B and C situated 16 m, 10 m and 2 m above datum. If the beginning of the pipe is connected to a tank which is filled with water to a height of 26 m above datum, find the discharge and the velocity and pressure heads at A, B and C. neglect all losses. Take atmospheric pressure equation to 10 m of water.

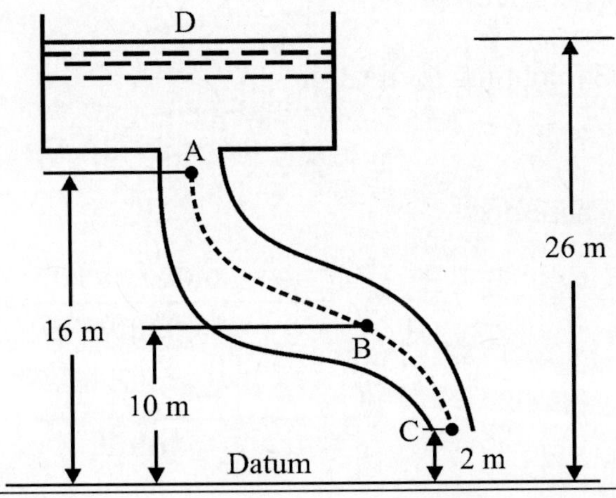

Fig. Ex. 5.10

Solution:

1) Cross sectional area at A, B, C 3000, 6000, 1250 mm^2.

2) Datum heads at A, B, C 16 m, 10 m and 2 m.

3) Head of water 26 m.

Applying Bernoulli's equation to point D, A, B and C in turns. Applying Bernoulli's equation to D and C:

$$\frac{P_D}{Pg} + \frac{v_D^2}{2\text{ g}} + Z_D = \frac{P_C}{Pg} + \frac{v_C^2}{2\text{ g}} + x_C$$

The point D and C are open to atmospheric, hence gauge pressure is zero.

$$0 + 0 + 26 = 0 + \frac{v_C^2}{2g} + 2$$

$$\frac{v_C^2}{2g} = 24$$

$$v_C^2 = 24 \times 2 \times 9.81 = 470.88$$

$$V_C = 21.7 \text{ m/s}$$

Therefore, $$Q = a_C\, v_C - \frac{1250}{10^6} \times 21.7 = 0.0271 \text{ m}^3\text{/s}$$

By continuity equation

$$Q = a_A\, v_A = a_B\, v_B = a_C\, v_C$$

Therefore,

$$V_A = \frac{Q}{a_A} = \frac{0.0271 \times 10^6}{3000} = 9.04 \text{ m/s}$$

$$V_B = \frac{Q}{a_B} = \frac{0.0271 \times 10^6}{6000} = 4.52 \text{ m/s}$$

i) Applying Bernoulli's theorem between water surface i.e. at D and A

$$\frac{P_D}{\rho G} + \frac{v_D^2}{2\,g} + Z_D = \frac{P_A}{\rho g} + \frac{v_A^2}{2\,g} + Z_A$$

$$0 + 0 + 26 = \frac{P_A}{\rho g} + \frac{(9.04)}{2\times 9.81} + 16$$

$$\frac{P_A}{\rho g} = 26 - 16 - 4.1652 = +\,5.8348 \text{ m of water}$$

$$= 15.8348 \text{ (absolute) m of water}$$

ii) Applying Bernoulli's theorem between D and B.

$$\frac{P_D}{\rho g} + \frac{v_D^2}{2g} + Z_D = \frac{P_B}{\rho g} + \frac{v_B^2}{2g} + Z_B$$

$$0 + 0 + 26 = \frac{P_3}{\rho g} + \frac{(4.52)^2}{2\times 9.81} + 10$$

$$\frac{P_B}{PB} = 26 - 10 - 1.0413 = 14.958 \text{ m of water gauge}$$

$$= 14.958 + 10 = 24.958 \text{ mm of water absolute}$$

Ex. 5.11: A 10 cm x 5 cm horizontal Venturimeter carries 50 lit/sec of water and has $C_d = 0.97$. Find the deflection in a mercury manometer connected between entrance and throat,

Solution:

$d_1 = 10$ cm $= 0.1$ m

$d_2 = 5$ cm $= 0.05$ m

$Z_1 = Z_2$ $\quad\because$ Venturimeter is horizontal

$Q = 50$ lit/sec. $= 50 \times 10^{-3}$ m^3/sec

$C_d = 0.97$

$S = 1.0$ (Flowing fluid is water)

$S_m = 13.6$ (Manometric fluid is mercury)

Using Bernoulli's equation at entrance and throat:

$$\frac{P_1}{Y} + \frac{V_1^2}{2g} + Z_1 = \frac{P_2}{\gamma} + \frac{V_2^2}{2g} + Z_2$$

$$\therefore \quad h = \left(\frac{P_1}{\gamma} + Z_1\right) - \left(\frac{p_2}{\gamma} + Z_2\right) = \frac{v_2^2}{2g} - \frac{v_1^2}{2g}$$

and

$$h = x\left(\frac{S_m}{S_a} - 1\right) = \frac{v_2^2}{2g} - \frac{v_1^2}{2g}$$

Using continuity equation (use of C_d) :

$$Q_{act}\ Q_{th} = C_d$$

$$Q_{act} = C_d \cdot Q_{th}$$

$$Q_{act} = C_d \cdot a_1 V_1 = C_d \cdot a_2 V_2$$

$$\therefore \quad 50 \times 10^{-3} = 0.97 \times \frac{\pi \times 0.1^2}{4} \times V_1$$

$$= V_1 = 6.56 \text{ m/sec}$$

Similarly

$$50 \times 10^{-3} = \frac{0.97 \times \pi 0.05^2}{4} V_2$$

$$\therefore \quad V_2 = 26.25 \text{ m/sec}$$

$$\therefore \quad x\left(\frac{13.6}{1.0} - 1\right) = \frac{26.25^3}{2 \times 9.81} - \frac{6.56^2}{2 \times 9.81}$$

$$\therefore \quad x = \frac{35.12 - 2.19}{12.6} = 2.61 \text{ m}$$

∴ Deflection in a mercury manometer is 2.61 m.

Ex. 5.12 :	A 30 m x 15 cm venturimeter is fitted in a horizontal pipeline carrying ail of specific gravity 0.8. The pressure at the inlet of the meter is 14 N/cm^2 and at the throat the vacuum pressure is 37 cm of mercury. Assuming 5% of the differential head is lost between the inlet and throat, determine the rate of flow of oil through the pipeline.

Solution:

d_1 30 cm = 0.3 m, d_2 = 15 cm = 0.15 m

S = 0.8, $P_1 = 14\ N/cm^2 = 14 \times 10^4\ N/m^2$

$\frac{P_2}{\gamma} = -37$ cm of mercury, $h_L = 0.05\left(\frac{P_1}{\gamma} - \frac{P_2}{\gamma}\right)$

$$\frac{P_2}{\gamma} = -37 \text{ cm of mercury}$$

$$= -0.37 \text{ cm of mercury}$$

$$= -0.37 \times 13.6 = -5.032 \text{ m of water}$$

$$= \frac{-5.032}{0.8} = -6.29 \text{ m of oil}$$

$$h_1 = 0.05\left(\frac{P_1}{\gamma} - \frac{P_2}{\gamma}\right)$$

$$= 0.05\left(\frac{14\times10^4}{9810\times0.8} - (-6.29)\right)$$

$$= -0.05\,(17.83 + 6.29)$$

$$= 1.206 \text{ m}$$

Horizontal Venturimeter

$\therefore$ $$Z_1 = Z_2$$

Using continuity equation:

$$a_1 V_1 = a_2 V_2$$

$$\frac{\pi d_1^2}{4} \times V_1 = \frac{\pi d_2^2}{4} \times V_2$$

$$0.0707\ V_1 = 0.01767\ V_2$$

	$\therefore \quad V_1 = 0.25\ V_2$ Using Bernoulli's equation: $\frac{P_1}{\gamma} + \frac{v_1^2}{2g} + Z_1 = \frac{P_2}{\gamma} + \frac{V_2^2}{2g} + Z_2 + h_L$ $\frac{14\times10^4}{0.8\times9810} + \frac{(0.25V_2)^2}{2\times9.81} = -6.29 + \frac{V_2^2}{2\times9.81} + 1.206$ $17.83 + 3.18 \times 10^{-3}\ V_1^2 = -6.29 + 0.05\ V_2^2 + 1.206$ $V_2 = 20.79$ m/sec Rate of flow of oil through the pipeline $Q = a_2V_2 = 0.1767 \times 20.79$ $= 3.67\ m^3/sec$

Ex. 5.13 :	A 500 mm diameter vertical Venturimeter discharges water through a throat of Q 250 mm diameter. The pressure difference measured by a inverted U tube manometer. The vertical distance between inlet and throat section is 500 mm. Determine (a) the difference in pressure between these two sections when the discharge through the meter is 600 litres/s and (*b*) the manometer deflection *x* as if the inverted U contains air. Take C_d as unity. **Solution:** $d_1 = 500$ mm $= 0.5$ m $\quad\quad d_2 = 250$ mm $= 0.25$ m

	Vertical distance $Z_1 - Z_2$ = 500 mm = 0.5 m

$C_4 = 1$

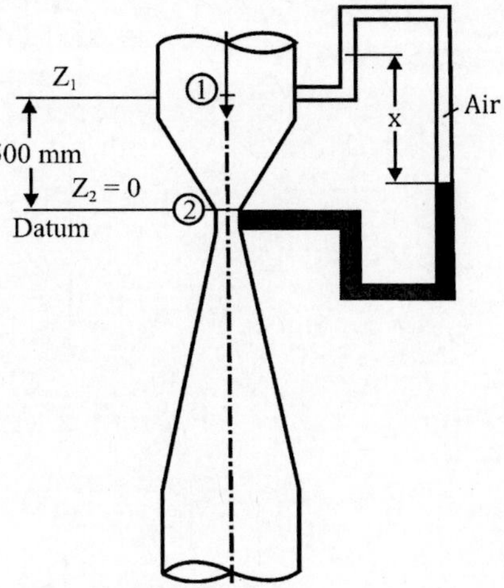

Fig. Ex. 5.13

$$Q = C_d \frac{a_1 a_2 \sqrt{2gh}}{\sqrt{a_1^2 - a_2^2}} = C_d \frac{a_1\sqrt{2gh}}{\sqrt{\left(\frac{a_1}{s_2}\right)^2 - 1}}$$

$$600 \times 10^{-3} = \frac{0.19635\sqrt{2\times 9.81\times h}}{\sqrt{16-1}}$$

$$600 \times 10^{-3} = 0.22456\sqrt{h}$$

$$\sqrt{h} = 2.67$$

$$h = 7.1389$$

$$h = \frac{P_1}{PE} = \frac{P_2}{Pg} + (Z_1 - Z_2)$$

$$7.1389 = \frac{P_1 - P_2}{Pg} + 0.5$$

$$\frac{P_1 - P_2}{Pg} = 7.1389 - 0.5 = 6.6389$$

	$\therefore \quad (P_1 - P_2) = \rho\, g \times 6.6389$ $= 1000 \times 9.81 \times 6.6389$ $= 65.1278 \times 10^3\ N/m^2$ $= 65.1278\ kPa$ $h = x\left[1 - \frac{S_{air}}{S_{water}}\right]$ $h = x\left[1 - \frac{\rho_{eir}}{\rho_{water}}\right]$ Taking $P_{air} = 1.2\ kg/m^3$ $P_{water} = 1000\ kg/m^3$ Therefore, $7.1389 = x\left[1 - \frac{1.2}{1000}\right]$ $7.1389 = x\,[0.999]$ $x = 7.146\ m$

Ex. 5.14 :	A venturimeter measures the flow of water in a 75 mm diameter pipe. The difference of head between the throat and the entrance of the meter is measured by a U-tube containing mercury, the mercury being in contact with the water. What should be the diameter of the throat of the meter in order that the difference in level of the mercury be 25 cm? When the quantity of water flowing in the pipe is 650 lpm. Assume 'C_d' of the meter as 0.97. **Solution:** $d_1 = 75\ mm = 0.075\ m \quad S_0 = 1.0$ $S_m = 13.6 \qquad x = 25\ cm - 0.25\ m$ $Q = 650\ lpm = \frac{650}{60}\ lps = \frac{650}{60} \times 10^{-3}\ m^3/sec = 0.01083\ m^3/sec$

$$C_d = 0.97$$

Using continuity equation:

$$Q = C_d\, a_1 V_1 = C_d\, a_2 V_2$$

$\therefore$ $$0.01083 = 0.97 \times 4.4179 \times 10^{-3} \times V_1$$

$\therefore$ $$V_1 = 2.53 \text{ m/sec}$$

Similarly $$0.01083 = 0.97 \times 0.7854\, d_2^2 \times V_2$$

$\therefore$ $$V_2 = \frac{0.0142}{d_2^2} \text{ m/sec}$$

Using Bernoulli's equation:

$$\frac{p_1}{\gamma} + \frac{v_1^2}{2g} + Z_1 = \frac{p_2}{\gamma} + \frac{V_2^2}{2g} + Z_2$$

$\therefore$ $$h = x\left(\frac{S_m}{S} - 1\right) = \left(\frac{P_1}{\gamma} + Z_1\right) - \left(\frac{P_2}{\gamma} + Z_2\right) - \frac{v_2^2}{2g} - \frac{v_1^2}{2\,g}$$

$\therefore$ $$x\left(\frac{S_m}{S_0} - 1\right) = \frac{V_2^2}{2g} - \frac{V_1^2}{2\,g}$$

$$0.25\left(\frac{13.6}{1.0} - 1\right)$$

$$= \left(\frac{0.0142}{d_2^2}\right)^2 \frac{1}{2\times 9.81} - \left(\frac{0.0142}{d_2^2}\right)^2 \frac{1}{2\times 9.81}$$

$$3.15 = \frac{1.0277\times 10^{-5}}{d_2^4} - 0.3262$$

$\therefore$ $$d_2 = \left(\frac{1.0277\times 10^{-5}}{3.15+0.3262}\right)^{1/4}$$

$$= (0.2956 \times 10{-}5)^{1/4}$$

$$= 0.041 \text{ m} = 41 \text{ mm}$$

Ex. 5.15 : A venturimeter whose inlet and throat diameter are 100 and 40 mm respectively is used to measure the flow of petrol (specific gravity = 0.78) along a pipe. The ends of a U tube containing mercury of specific gravity 13.6 are connected to the meter at the inlet and throat, and the difference of levels is observed to be 480 mm when the discharge is 15.6 litres/s. Determine the theoretical venturihead (neglecting friction) and hence deduce the coefficient of the meter.

Assuming that friction losses are directly proportional to the measured venturihead, determine this head in mm of mercury when the discharge is halved.

Solution:

Data : $d_1 = 100$ mm $= 0.1$ m $\qquad d_2 = 40$ mm $= 0.04$ m

Specific gravity of petrol = 0.78

$Q = 15.6$ lit/sec $= 15.6 \times 10^{-3}$ m^3/sec

$x = 480$ mm of mercury $= 0.480$ m

Neglecting Losses:

$$\text{Venturihead} = \frac{p_1 - p_2}{\rho g \text{ petrol}} + (Z_1 - Z_2) = \frac{V_2^2 - v_1^2}{2\,g} = H$$

$$= x\left(\frac{S_m}{S} - 1\right) - \frac{480}{1000}\left(\frac{13.6}{0.78} - 1\right) - 7.889 \text{ m}$$

$$\therefore \qquad \frac{v_2^2 - v_1^2}{2g} = 7.889$$

Using continuity equation:

$$a_1 V_1 = a_2 V_2$$

$$(0.1)^2\, V_1 = (0.04)^2\, V_2$$

$$6.25\, V_1 = V_2$$

Substituting in step (1)

$$\frac{v_2^2 - \frac{v_2^2}{(6.25)^2}}{2\text{ g}} = 7.889$$

$$\frac{v_2^2}{2\text{ g}}\left[\frac{38.0625}{39.0625}\right] = 7.889$$

$$v_2^2 = 158.848$$

$$v_2 = 12.6 \text{ m/s}$$

$$Q = a_2 V_2 = \frac{\pi}{4}(0.04)^2 \times 12.6$$

$$= 0.015838 \text{ m}^3\text{/s} - 15.838 \text{ litres/s}$$

$$C_d = \frac{Q_{actual}}{Q_{theoretical}} = \frac{15.6}{15.838} = 0.9849$$

As per question $h_f \propto$ vesturihend :

$$Q = \frac{15.6}{2} = 7.8 \text{ litres/s} - 7.8 \times 10^{-3} \text{ m}^3\text{/s}$$

$$\left(\frac{P_1}{\rho_{Petrol}\text{ g}} + Z_1\right) - \left(\frac{P_2}{\rho_{Petrol}\text{ g}} + Z_2\right) - h_f = \frac{V_2^2 - V_1^2}{2\text{g}}$$

$$h - h_f = \frac{V_2^2 - V_1^2}{2\text{g}}$$

or, $$x\left(\frac{S_m}{S} - 1\right) - x = \frac{V_2^2 - V_1^2}{2\text{g}}$$

or, $$x\left[\frac{13.6}{0.78} - 1\right] - x = \frac{V_2^2 - V_1^2}{2\text{g}}$$

or, $$16.7458\, x - x = \frac{V_2^2 - V_1^2}{2\text{g}}$$

or, $$15.4358\, x - x = \frac{V_2^2 - V_1^2}{2\text{g}}$$

	Using continuity equation: $$V_1 = \frac{Q}{a_1} = \frac{7.8\times10^{-3}}{\frac{\pi}{4}(0.1)^2} = 0.993 \text{ m/s}$$ $$V_2 = 6.25\ V_1 - 6.207 \text{ m/s}$$

Ex. 5.16 :	If in the venturimeter the quantity of water passing through the meter will only be proportional to the measured venturihead h show that the head lost in friction h_l is, proportional to the head difference h_V due to increased velocity. A venturimeter has a coefficient of discharge of 0.98 and the frictional loss in the diverging cone is twice that in the converging cone. What will be total head lost in friction in the meter when the measured difference head is equivalent to 500 m of water? **Solution:** **Given :** Q ∝ venturihead h, $C_d = 0.98$ $$Q_{act} = C_d \frac{a_1\sqrt{2gh}}{\sqrt{\left(\frac{a_1}{a_2}\right)^2 - 1}} \quad \text{---(1)}$$ If frictional losses are used $$Q_{act} = \frac{a_1\sqrt{2g(h-h_f)}}{\sqrt{\left(\frac{a_1}{a_2}\right)^2 - 1}} \quad \text{---(2)}$$ $$= a_1\sqrt{\frac{2gh_v}{\left(\frac{a_1}{a_2}\right)^2} - 1} \quad \text{where } h - h_f = h_v$$ Equate Equations (1) and (2) :

	$\therefore \qquad C_d = \sqrt{\dfrac{h_v}{h}}$ ---(3) $\therefore \qquad C_d = \sqrt{\dfrac{h_v}{h_v+h_t}}$ ---(4) If Q is proportional to $\sqrt{h}$ then by step (1) C_d is constant Now suppose $\qquad k = \dfrac{h_v}{h_v+h_f}$ $\qquad$ where k is constant. $\therefore \qquad h_f = \dfrac{1-k}{k} hv$ $\therefore$ $h_f \propto h_v$ proved From Equations (3) & (4) $h_f = h(1 - C_d^2)$ = Loss of head in converging cone. If head loss in diverging cone $= 2h_f$ Total head loss $= h_f + 2\,h_f - 3\,h_f$ $\therefore \qquad 3\,h(1 - C_d^2) = 3 \times 0.5[1 - 0.98^2]$ $= 0.0594$ m of water

Ex. 5.17 :	(a) Coefficient of discharge for a venturimeter used for measuring the discharge of an incompressible fluid was found to be constant provided that the rate of flow exceeds a certain value. Show that water under these conditions the loss of head in the convergent portion of the venturi can be expressed by kQ^2 meter where k is a constant and Q is the rate of flow in m^3/s. (b) A venturimeter with 75 mm diameter throat is installed in a 150 mm diameter pipeline. The pressure at the entrance to the meter is

	0.7 bar gauge and it is undesirable that the pressure should, at any point, fall below 66 kPa absolute. Assuming C_d for the meter as 0.96, find the maximum flow for which it may be used. Take the relative density of the liquid as 0.75 and atmospheric pressure 750 mm of mercury. (10 Marks) **Solution:** d_2 = 75 mm = 0.075 m, d_1 = 150 mm = 0.150 m P_1 = 0.7 bar gauge (if 750 mm of mercury is equal to 1 bar) $P_1 - 1 + 0.7 = 1.7$ bar absolute Relative density of liquid = 0.75, $P_2 = 56$ kPa $= 56 \times 10^3$ Pa Applying the Bernoulli's equation between points 1 and 2: $$\frac{P_1}{\rho g} + \frac{V_1^2}{2g} + Z_1 = \frac{p_2}{\rho g} + \frac{V_2^2}{2g} + Z_1 + h_f$$ $$\left(\frac{P_1}{\rho g} + Z_1\right) - \left(\frac{P_2}{\rho g} + Z_2\right) - h_f = \frac{v_2^2}{2g} - \frac{v_1^2}{2g}$$ $$V_2 = \frac{a_1\sqrt{2g(h-h_f)}}{\sqrt{a_1^2 - a_2^2}}$$ or $$h - h_f = \frac{V_2^2 - V_1^2}{2g} g$$ $$Q_{act} = a_2 V_2 = \frac{a_1 a_2 \sqrt{2g(h-h_f)}}{\sqrt{a_1^2 - a_2^2}}$$ when head lost due to friction is not included in the equation, the equation is known as an ideal equation from which we will get the theoretical discharge (Q_{th}).

$$Q_{th} = \frac{a_1 a_2 \sqrt{2g}}{\sqrt{a_1^2 - a_2^2}}$$

Using the coefficient of discharge

$$Q_{act} = C_d\, Q_{th}$$

$$= C_d \frac{a_1 z_2 \sqrt{2g}}{\sqrt{a_1^2 - a_2^2}}$$

Equating Equations (2) and (3)

$$h - h_f = C_d^2 h$$

$$h = \frac{h_f}{1 - C_d^2}$$

Substituting this value of h in Equation (3)

$$Q_{act} = C_d \frac{a_1 a_2 \sqrt{h_f(1 - c_d^2)}}{\sqrt{a_1^2 - a_2^2}}$$

$$Q = \frac{C_d a_1 a_2 \sqrt{2gh}}{\sqrt{(1 - c_d^2)}\sqrt{a_1^2 - a_2^2}}$$

Squaring on both the sides

$$Q^2 = \left(\frac{C_d^2}{1 - C_d^2}\right) \frac{a_1^2 a_2^2 2gh}{(a_1^2 - a_2^2)}$$

If C_d is constant then

$$h_f = kQ^2 \text{ proved}$$

where $$k = \left(\frac{1 - C_d^2}{C_d^2}\right)\left(\frac{a_1^2 - a_2^2}{a_1^2 a_2^2 2gh}\right)$$

	$h_1 = \frac{P_1}{(\text{sp. gr.})\rho g} = \frac{1.7\times10^5}{0.75\times1000\times9.81} = 23.1$ m absolute $h_2 = \frac{P_2}{(\text{sp. gr.})\rho g} = \frac{56\times10^3:}{0.75\times1000\times9.81} = 7.61$ m absolute
∴	$h_1 - h_2$ = difference in absolute reading = 15.49 m $a_1 = \frac{\pi}{4} \times (0.15)^2 = 0.01767\ m^2$
and	$\left(\frac{a_1}{a_2}\right) = 4$ $Q = \frac{C_d a\sqrt{2gh}}{\sqrt{\left(\frac{a_1}{a_2}\right)^2 - 1}} = \frac{0.96\times0.01767\sqrt{2\times9.81\times15.49}}{\sqrt{16-1}}$ $= 0.07636\ m^3/s$

Ex. 5.18 :	A 15 cm x 5 cm venturimeter is provided in a vertical pipe carrying crude oil (specific gravity = 0.8). The flow is in upward direction. The difference of elevation between the entrance and throat section of the venturimeter is 25 cm. The difference in level between the two limbs of U-tube mercury manometer recorded is 20 cm. Calculate (i) Flow rate of oil (ii) Pressure difference between the entrance and the throat section. Take coefficient discharge as 0.95. **Solution:** $d_1 = 0.15$ m, $d_2 = 0.05$ m, $S = 0.8$ Sm = 13.6 $Z_1 - Z_2 = 0.25$ m, $x = 0.2$ m, C_d = 0.95

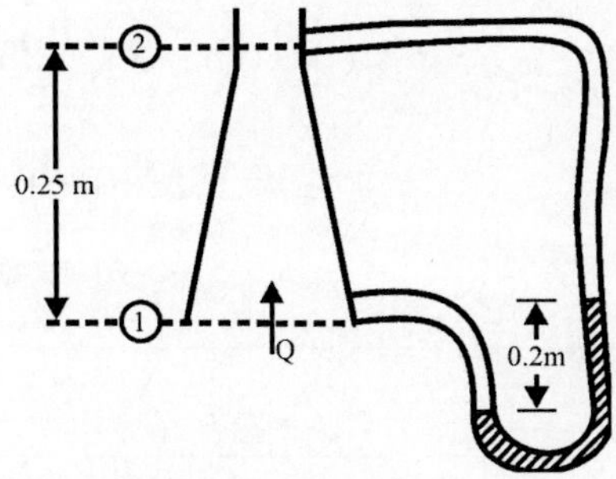

Fig. Ex. 5.18

Piezometric head difference

$$h = \frac{v_2^2 - v_1^2}{2g} = \left(\frac{P_1}{\gamma} + Z_1\right) - \left(\frac{P_2}{\gamma} + Z_2\right)$$

$$a_1 = \frac{\pi}{4} \text{ x } (0.15)^2 = 0.0177 \text{ m}^2$$

$$a_2 = \frac{\pi}{4} \text{ x } (0.05)^2 = 1.963 \text{ x } 10^{-3} \text{ m}^2$$

$$h = x\left(\frac{S_m}{s} - 1\right) = 0.2\left(\frac{13.6}{0.8} - 1\right) - 3.2 \text{ m}$$

$$0.0177 \text{ V}_1 = 1.963 \text{ x } 10^{-3} \text{ V}_2$$

$$\therefore \quad V_1 = 0.11 \text{ V}_2$$

$$h_1 = \frac{V_2^2 - 0.11 \text{ V}_2^2}{2 \text{ g}}$$

$$V_2 = 7.97 \text{ m/sec}$$

By using relation in step 1:

$$Q_{act} = C_d \cdot a_2 \text{ V}_2$$

$$= 0.95 \text{ x } 1.963 \text{ x } 10^{-3} \text{ x } 7.97$$

$$= 0.0149 \text{ m}^3\text{/sec}$$

$$h = \left(\frac{P_1}{\gamma} + Z_1\right) - \left(\frac{P_2}{\gamma} + Z_2\right)$$

	$\therefore \quad \frac{P_1}{\gamma} - \frac{P_2}{\gamma} = h - Z_1 + Z_2$ $= 3.2 - 0 + 0.25$ $= 3.45$ m of oil

Ex. 5.19:	A venturimeter with a throat diameter of 10 cm is connected to a 20 cm diameter main carrying water. The head loss between the inlet and the throat diameter main carrying water, The head loss between the inlet and the throat is known to be $0.1\frac{V_2^2}{2g}$ where V_2 is the velocity at the throat. Estimate the discharge when the inverted differential U-tube manometer connected to the inlet and throat records a reading of 30 cm. The manomatric fluid has a relative density of 0.75. What is the coefficient of discharge of the meter?

Solution:

$d_1 = 0.2$ m, $\quad d_2 = 0.1$ m, $\quad h_L = 0.1\frac{v_2^2}{2\,g}$

$x = 0.3$ m, $\quad Sm = 0.75$, $\quad S = 1.0$

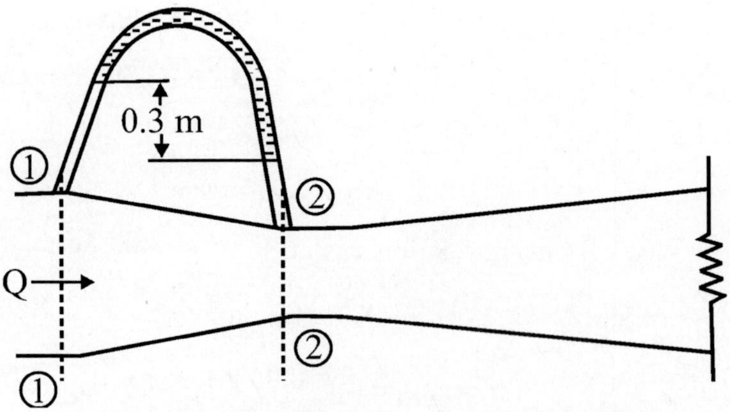

Fig. Ex. 5.19

By continuity equation:

$$a_1V_1 = a_2V_2$$

$$\frac{\pi}{4} \text{ x } 0.2^2 \text{ x } V_1 = \frac{\pi}{4} \text{ x } 0.1^2 \text{ x } V_2$$

$$\therefore \quad V_1 = 0.25\ V_2$$

By Bernoulli's theorem:

$$\frac{P_1}{\gamma} + \frac{V_1^2}{2g} + Z_1 = \frac{P_2}{\gamma} + \frac{V_2^2}{2g} + Z_2 + h_L$$

$$\left(\frac{P_1}{\rho g} + Z_1\right) - \left(\frac{P_2}{\rho g} + Z_2\right) = \frac{v_2^2 - v_1^2}{2g} + h_L$$

$$h = \frac{v_2^2 - v_1^2}{2g} + h_L$$

$$x\left(1 - \frac{S_m}{s^2}\right) = \frac{V_2^2 - v_1^2}{2\ g} + h_L$$

$$0.3\left(1 - \frac{0.75}{1.0}\right) = \frac{V_2^2 - v_1^2}{2\ g} + h_L$$

$$0.3\left(1 - \frac{0.75}{1.0}\right) = \frac{V_2^2}{2\ g} - \frac{(0.25\ V_2)^2}{2\ g}\ (0.9)$$

$$0.075 = V_2^2\left(\frac{1 - 0.5625}{2 \times 9.81}\right)$$

$$V_2 = 1.25 \text{ m/s}$$

By continuity equation $Q = a_1V_1 = a_2V_2$

$$\therefore \quad Q = \frac{\pi}{4} \text{ x } 0.12 \text{ x } 1.25$$

$$= 9.81 \text{ x } 10^{-3} \text{ m}^3/\text{s}$$

$$h_L = \frac{0.1 v_2^2}{2g} = \frac{0.1 \times 1.25^2}{2 \times 9.81}$$

	$= 7.9638 \times 10^{-3}$ m $= x\left(1-\frac{S_m}{s}\right) = 0.3\left(1-\frac{0.75}{1.0}\right)$ $= 0.075$ m $C_d = \sqrt{\frac{h_L - h}{h}}$ $= 0.94$

Ex. 5.20:	Water flows at the rate of 10.5 liters/s through a 150 mm diameter pipe in 0 which an orifice meter with a 100 mm diameter orifice is fitted. If the pressure drop across the meter is recorded as a 18 mm difference in levels of mercury in a U tube manometer, what would be the coefficient of discharge C_d? Assume value of C_V. If the orifice were 125 mm diameter what would be the head loss in m of water, for the above values of Q, C_d and C_c? Also calculate the pressure drop across the meter recorded in U tube manometer. **Solution:** $Q = 10.5$ lit/sec $= 10.5 \times 10^{-3}$ m³/s $d_1 = 0.15$ m, $d_2 = 0.10$ m, $x = 18$ mm mercury $h = x\left(\frac{S_m}{S} - 1\right) = \frac{18}{1000}\left(\frac{13.6}{1} - 1\right)$ $= 0.2268$ m $Q = \frac{C_d \cdot a\sqrt{2gh}}{\sqrt{1 - C_e^2\left(\frac{a}{A}\right)^2}}$

Assume

$$C_V = 0.95$$

$$C_d = C_c \cdot C_V$$

$$\therefore \quad C_c = \frac{C_d}{0.95}$$

$$\therefore \quad 10.5 \times 10^{-3} = \frac{C_d \times \frac{\pi}{4}(0.15)^2\sqrt{2\times 9.81\times 0.2268}}{\sqrt{1-\left(\frac{C_d}{0.95}\right)^2\left(\frac{0.10}{0.15}\right)^2}}$$

$$10.5 \times 10^{-3}\sqrt{1-\left(\frac{C_d}{0.95}\right)^2(0.444)} = C_d \times (0.0176)\sqrt{4.445}$$

$$\frac{\sqrt{1-\left(\frac{C_d}{0.95}\right)^2(0.444)}}{C_d} = 3.53$$

$$\frac{1-0.492C_d^2}{C_d} = 12.46$$

$$? + 0.492\, C_d^2 + 12.46\, C_d - 1 = 0$$

$$\therefore \quad C_d = 0.6076$$

$$\therefore \quad C_d = \frac{C_d}{0.95} = 0.6396$$

Using values of C_d and C_c with 0.125 mm diameter orifice the equation of discharge:

$$Q = \frac{C_d \cdot a\sqrt{2gh}}{\sqrt{1-c_e^2\left(\frac{a}{A}\right)^2}}$$

$$10.5 \times 10^{-3} = \frac{0.6076 + \frac{\pi}{4}\times(0.15)^2\sqrt{2\times 9.81\times h}}{\sqrt{1-(0.6396)^2\left(\frac{0.125}{0.150}\right)^2}}$$

	Solving the above equation: $\sqrt{h} = 0.2848$ h = 0.081 m $h = 12.6 \times x$ h = (0.081)12.6 $x = 6.437 \times 10^{-3} = 6.437$ mm

Ex. 5.21 :	An orifice meter of orifice diameter 10 cm is inserted in a pipe of 20 cm (1) diameter. The pressure gauge fitted on upstream and downstream give readings of 20 N/cm^2 and 10 N/cm^2 respectively. $C_d = 0.6$. Find the discharge. **Solution:** $d_1 = 20$ cm = 0.2 m, $d_2 = 10$ cm = 0.1 m $P_1 = 20$ N/cm^2 = 20×10^4 N/cm^2, $P_2 = 10$ N/cm^2 = 10×10^6 N/cm^2, $C_d = 0.6$ $a_1 = \frac{\pi}{4} \times d_1^2 = \frac{\pi}{4} \times 0.2^2 = 0.0314 \text{ m}^2$ $a_2 = \frac{\pi}{4} \times d_2^2 = \frac{\pi}{4} \times 0.1^2 = 0.00785 \text{ m}^2$ $Q = a_1V_1 = a_2V_2$ $= 0.0314\ V_1 = 0.00785\ V_2$ $V_1 = 0.25\ V_2$ By Bernoulli's theorem: $\frac{P_1}{\gamma} + \frac{V_1^2}{2g} + Z_1 = \frac{P_2}{\gamma} + \frac{V_2^2}{2g} + Z_2$

$$\frac{20\times10^4}{9810}+\frac{(0.25\,V_2)^2}{2\times9.81}+0=\frac{10\times10^4}{9810}+\frac{V_2^2}{2\times9.81}$$

$$20.387 + 3.185 \times 10^{-3}\, V_2^2 = 10.194 + 0.051\, V_2^2$$

$$V_2 = 14.60 \text{ m/sec}$$

$$Q = C_d \cdot a_2V_2 = 0.6 \times 0.00785 \times 14.60$$

$$= 0.06878 \text{ m}^3/\text{sec}$$

Ex. 5.22 : A pitot-static tube is used to measure the velocity of water in a pipe. The (5) stagnation pressure head is 6 m and static pressure head is 5 m. Calculate the velocity of flow assuming the co-efficient of tube equal to 0.98.

Solution:

Stagnation pressure head = 6 m,

Static pressure head = 5 m

$\therefore$ h = 6 – 5 = 1 m

Using equation of velocity using pitot tube find velocity.

$$V = C_V\sqrt{2gh}$$

$$= 0.98\sqrt{2 \times 9.81 \times 1}$$

$$= 4.34 \text{ m/s}$$

Ex. 5.23 : A sub-marine moves horizontally in sea and has its axis 15 m below the surface of water. A pitot-tube properly placed just in front of the sub-marine and along its axis is connected to the two limbs of a U-tube containing mercury. The difference of mercury level is found to be 170 mm. Find the speed of the sub-marine knowing that the specific gravity of mercury is 13.6 and that of sea-water is 1.026 with respect of fresh water.

	Solution: x = 170 mm = 0.17 m Specific gravity of mercury = 13.6 Specific gravity of sea-water = 1.026 $$h = x\left[\frac{S_m}{S} - 1\right]$$ $$= 0.17\left[\frac{13.6}{1.026} - 1\right]$$ $$= 2.0834 \text{ m}$$ $$V = \sqrt{2gh} = \sqrt{2 \times 9.81 \times 2.0534}$$ $$= 6.393 \text{ m/s}$$

Ex. 5.24:	A pitot-tube is inserted in a pipe of 300 mm diameter. The static pressure in pipe is 100 mm of mercury (vacuum). The stagnation pressure at the centre of the pipe, recorded by the pitot-tube is 0.981 N/cm^2. Calculate the rate of flow of water through pipe, if the mean velocity of flow is 0.85 times the central velocity. Take C_V = 0.98. **Solution:** d = 0.3 m Static pressure = 100 mm of mercury (vacuum) Stagnation pressure = 0.981 N/cm^2 $$V_{mean} = 0.85\ V_{central}$$ $$C_V = 0.98$$ Static pressure 100 mm of mercury (vacuum) $$= \frac{-100}{1000} \times 13.6$$ $$= -13.6 \text{ m of water}$$

	$= 0.981\ N/cm^2$ $= 0.981 \times 104\ N/m^2$ Stagnation pressure $= 0.981\ N/cm^2$ Stagnation pressure head $= \frac{0.981\times10^4}{1000\times9.81} = 1$ m $\therefore$ h = Stagnation pressure head – Static pressure head $= 1 - (-1.36)$ $= 2.36$ m of water Velocity at centre $= C_V = \sqrt{2gh}$ $= 0.98\sqrt{2 \times 9.81 \times 2.36} - 6.668$ m/s Mean velocity $= 0.85 \times 6.668 = 5.667$ m/s $Q = a\,V_{mean} = 5.6678 \times 0.07068 = 0.4006\ m^3/sec.$

Ex. 5.25 :	An orifice is located in the side of a tank which issues oil of relative density (T) 0.85 under a pressure of 0.2 kg/cm² at a rate of 4.5 *l*/s. The diameter of orifice at vena contracta is 3 cm. Find coefficient of velocity **Solution:** Pressure $0.2\ kg/cm^2 = 2000\ kg/m^2$ Pressure head $= \frac{P}{\gamma} = \frac{2000}{0.85\times1000}$ $H = 2.353$ m Discharge $Q = 4.5/s = 4.5\ 10^{-3}$ m3/sec Diameter at V.C. = 3 cm = 0.03 m Area of jet at V.C. $= \frac{m}{4} \times 0.03^2$

$= 7.068 \times 10^{-4}\ m^2$

Velocity at V.C. = Q / a_c

$$= \frac{4.5\times10^{-3}}{7.068\times10^{-4}} = 6.366 \text{ m/s}$$

$$C_4 = \frac{V}{\sqrt{2\,gH}}$$

$$= \frac{6.366}{\sqrt{2\times9.81\times2353}} = 0.937$$

Ex. 5.26: A tank has two identical orifices in one of its vertical sides. The upper orifice is 3.2 m below the water surface and lower one is 5.5 m below the water 5 surface. If value of C_v for each orifice is 0.96, find the point of intersection of two jets.

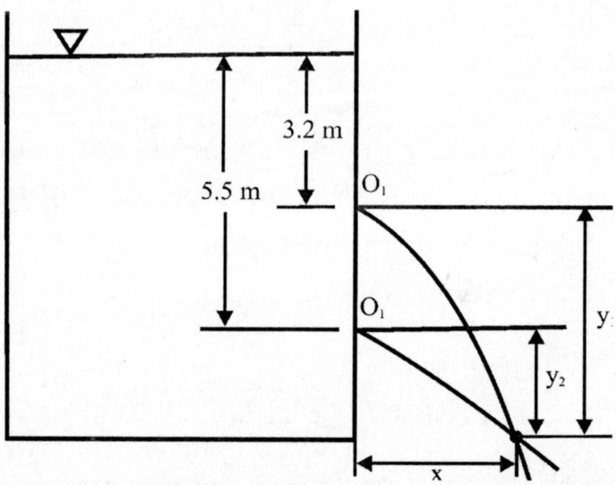

Fig. Ex. 5.26

Solution:

$C_v = 0.96$

Positions of orifices as shown in Fig. Ex. 5.26

Let the jets intersect at point B.

From Fig. Ex. 5.26

$H_1 = 5.5$ and $H_2 = 3.2$

and $$y_1 = y_2 + (5.5 - 3.2)$$

$$y_1 = y_2 + 23$$

As both the orifices have same C_y

$$C_{v_1} = C_{v_2}$$

$$\sqrt{\frac{x^2}{4H_1y_1}} = \sqrt{\frac{x^2}{4H_2y_2}}$$

$$\sqrt{\frac{x^2}{4\times 5.5\times y_1}} = \sqrt{\frac{x^2}{4\times 3.2\times y_2}}$$

∴ $$\frac{x^2}{4\times 5.5\times y_1} = \frac{x^2}{4\times 3.2\times y_2}$$

∴ $$3.2\ y_1 = 5.5\ y_2$$

But $$y_1 = y_2 + 23$$

∴ $$3.2\ (y_2 + 23) = 5.5y_2$$

∴ $$y_2 = 3.2$$

$$C_v = \sqrt{\frac{x^2}{4H_2y_2}}$$

$$96 = \sqrt{\frac{x^2}{4\times 5.5\times 3.2}}$$

$$x = 8.055\ \text{m}$$

Ex. 5.27:	Compensation water is to be discharged by two circular orifices of the same diameter situated at the bottom of a vessel having a constant head of 2m. If the demand is 18×10^5 lit/day what diameter

will be required for each orifice? Take coefficient of contraction and velocity as 0.62 and 0.98 respectively.

Solution:

$H = 2m$

Demand of water 18×10^5 lit/day

$$C_c = 0.62, \quad C_v = 0.98$$

$$C_d = C_c \times C_v$$

$$= 0.98 \times 0.62$$

$$C_d = 0.6076$$

$$Q = 18000000 \text{ lit/day}$$

$$= 18000 \text{ m}^3\text{/sec}$$

$$= \frac{18000}{24 \times 60 \times 60} \text{ m}^3\text{/sec}$$

$$Q_{act} = 0.2083 \text{ m}^3\text{/sec}$$

$$C_d = \frac{Q_{act}}{Q_{th}}$$

$\therefore$

$$Q_{th} = 0.6076 \times 0.2083$$

$$= 0.3428 \text{ m}^3\text{/sec}$$

$$V_{th} = \sqrt{2gH}$$

$$= \sqrt{2 \times 9.81 \times 2}$$

$$= 6.264 \text{ m/s}$$

Theoretical discharge through each orifice

$$= \frac{0.3428}{2} - 0.1714 \text{ m}^3\text{/sec}$$

$$\text{Area of orifice} = \frac{Q_{th}}{V_{th}}$$

$$= \frac{0.1714}{6.264}$$

	$a = 0.02736\ m^2$ $\therefore \quad \frac{\pi}{4} \times d^2 = 0.02736$ $d = 0.1866\ m$ $= 18.66\ cm$

Ex. 5.28 :	Water flows through a uniform diameter pipe of 200 mm. Points A and B are at elevations of 6 m and 8 m respectively along the inclined pipe. Pressures at A and B are 50 kPa and 20 kPa respectively. If rate of flow is 60 lit/sec. Determine : (1) Direction of flow (2) Head loss between these points (Dec. 98, 4 Marks) **Solution:** $D = 200\ mm = 0.2\ m \quad Z_A = 6\ m \quad Z_B = 8\ m$ $P_A = 50\ kPa \quad P_B = 20\ kPa \quad Q = 60 \times 10^{-3}\ m^3/sec$ $V_A = V_B - \frac{Q}{A} = \frac{60\times10^{-3}}{\pi/4\times0.2^2}$ $= 1.91\ m/s$ Velocity head, $\frac{V_A^2}{2g} = \frac{V_B^2}{2g} = \frac{(1.91)^2}{2\times9.81}$ $= 0.186\ m$ Total head at A $= \frac{P}{\gamma} + \frac{V^2}{2g} + Z$ $= \frac{50}{9.81} + 0.186 + 6$ $= 11.283$ m of water

$$\text{Total head B} = \frac{P}{\gamma} + \frac{V^2}{2\,g} + Z$$

$$= \frac{20}{9.81} + 0.186 + 8$$

$$= 10.225 \text{ m}$$

∴ Water flows from A to B.

$$\text{Head loss} = \text{Total energy at A} - \text{Total energy at B}$$

$$= 1.058 \text{ m}$$

Ex. 5.29 : A horizontal venturimeter of specification 200 x 100 mm is used to measure the discharge of an oil of specific gravity 0.8. A mercury manometer is used for the purpose. If the discharge is 100 lit/sec and if the coefficient of discharge of meter is 0.98, find the deflection of the manometer.

Solution:

D = 200 mm = 0.2 m d = 100 mm = 0.1 m

Specific gravity = 0.8 $Q = 100 \times 10^{-3}$ m^3/sec $C_d = 0.98$

$$Q = C_d \sqrt{h}$$

$$C = \frac{a_1\sqrt{2g}}{\sqrt{\left(\frac{d_1}{d_2}\right)^4 - 1}}$$

$$a_1 = \frac{\pi}{4}(0.2)^2 = 0.0314 \text{ m}^2$$

$$\left(\frac{d_1}{d_2}\right) = 2$$

$$\therefore \quad C = \frac{0.0314\sqrt{2\times 9.81}}{\sqrt{(2)^2 - 1}} = 0.036 \text{ m}^{5/2}/\text{sec}$$

$$\therefore \quad Q = C \cdot C_d\sqrt{h}$$

	$0.1 = 0.98 \times 0.036\sqrt{h}$
	h = 8.066 m
But	$h = \left(\frac{S_m}{50} - 1\right)x$
	$8.066 = \left(\frac{13.6}{50} - 1\right)x$
Manometric deflection	$x = 50.4$ cm

UNIT SUMMARY

1 The study of fluid motion with the forces causing the flow is called as dynamics of fluid flow, which is analysed by the Newton's second law of motion.

2 Bernoulli's equation is obtained by integrating the Euler's equation of motion states:

For a steady, ideal flow of an incompressible fluid, the total energy which consists of pressure energy, kinetic energy and datum energy at any point of the fluid is constant.

Mathematically, Bernoulli's equation.

$$\frac{P_1}{\rho g} + \frac{V_1^2}{2g} + Z_1 = \frac{P_2}{\gamma} + \frac{V_2^2}{2g} + Z_2$$

where $\frac{p}{\rho g}$ = Pressure energy per unit weight = Pressure head

$\frac{V^2}{2\,g}$ = Kinetic energy per unit weight = Velocity head $Q = C \cdot C_d\sqrt{h}$

Z = Datum energy per unit weight = Datum head

3. Bernoulli's equation for real fluids

$$\frac{P_1}{\rho g} + \frac{V_1^2}{2g} + Z_1 = \frac{P_2}{\gamma} + \frac{V_2^2}{2g} + Z_2 + h_L$$

Where h_L = loss of head between section 1 and 2.

4. The kinetic energy correction factor is given by,

$$\alpha = \frac{1}{A}\int \left(\frac{V}{V}\right)^3 da$$

5 The Bernoulli's equation with the kinetic energy correction factor is,

$$\frac{P_1}{\rho g} + \frac{\alpha V_1^2}{2g} + Z_1 = \frac{P_2}{\rho g} + \frac{\alpha V_2^2}{2g} + Z_2$$

6 The momentum equation states that the net force acting on a fluid mass is equal to the change in momentum per second in that direction.

This is given by $F = \frac{d}{dt}(mV)$

The impulse momentum equation is given by

$$F \cdot dt = d\,(mV)$$

7 The modified form of Impulse momentum equation is,

$$F = \rho Q\,(V_1 - V_2)$$

8 The momentum correction factor β is given by,

$$\beta = \frac{1}{A}\int \left(\frac{v}{V}\right)^2 \cdot dA$$

9 The discharge Q, through a venturimeter is given by,

$$Q = C_d\,\frac{a_1 a_2\sqrt{2gh}}{\sqrt{a_1^2 - a_2^2}}$$

Where a_1 = Area at inlet of venturimeter

a_2 = Area at throat of venturimeter

C_d = Coefficient of discharge

H=Difference of pressure head in terms of fluid flowing through venturimeter.

10. For measurement of pressure/piezometric head difference if U tube manometer is used.

$$h = x\left(\frac{S_m}{S} - 1\right) \qquad \text{where } S_m > S$$

Where S_m is the specific gravity of manometric liquid.

S is the specific gravity of liquid flowing through pipe.

x is the deflection in terms of manometric liquid and

$$h = x\left(1 - \frac{S_m}{S}\right) \qquad \text{where } S_m < S$$

11. Discharge through orificemeter is expressed by

$$Q = C_d \cdot C \cdot a\sqrt{2gh}$$

where C_d is coefficient of discharge.

a is cross sectional area of the orifice.

h is the head difference between section of the pipeline and Vena Contracta.

C is constant of venturimeter, can be expressed by,

$$\frac{1}{\sqrt{1 - C_c^2\left(\frac{a}{A}\right)^2}}$$

12. Pitot tube is used to find the velocity of a flowing fluid at any point in a pipe or channel. The velocity is given by the relation,

$$V = C_v\sqrt{2gh}$$

Where C_v is coefficient of velocity of pitot tube.

h - rise of liquid in the liquid representing dynamic pressure head for pipes and rise of liquid in the tubes above free surface of liquids for channels.

13. The discharge through triangular notch can be expressed as,

$$Q = \frac{8}{15} C_d\sqrt{2\,g}\tan\frac{\theta}{2} H^{5/2}$$

Where C_d is coefficient of discharge.

θ is the angle of notch.

H is the head above crest level.

14. With consideration of velocity of approach the expression for discharge through triangular notch is given by,

$$Q = \frac{8}{15} C_d\sqrt{2\,g}\tan\frac{\theta}{2}\left[(H + ha)^{5/2} - ha^{5/2}\right]$$

where ha is the additional head introduced due to velocity of approach.

$$ha = \frac{v_a^2}{2g}$$

5.12: Exercise

5.12.1: Objective Questions:

1. Which forces are neglected to obtain Euler's equation of motion from Newton's second law of motion?
   a) Viscous force, Gravity force, Turbulence force
   d) Viscous force, Turbulence, Turbulence force, Compressible force
   b) Gravity force, Turbulence force, Compressible force
   c) Body force e force, Body force
   Ans: (a)

2. Navier-Stoke's equation can be obtained from Reynolds's equation by not considering which type of force?
   a) Turbulence force b) Gravity force
   c) Compressible force d) Viscous force
   Ans: (a)

3. Which of the following assumption is incorrect in the derivation of Bernoulli's equation?
   a) The fluid is ideal b) The flow is steady
   c) The flow is incompressible d) The flow is rotational
   Ans: (d)

4. Cheapest device for measuring fluid flow / discharge rate.
   a) Venturimeter b) Pitot tube
   c) Orificemeter d) None of the mentioned
   Ans: (c)

5. If in a fluid, while applying Newton's second law of motion, compressibility force is neglected then what equation is obtained?
   a) Navier Stoke's Equation
   b) Reynold's equation of motion
   c) Euler's Equation of motion
   d) Continuity Equation for fluid flow

   Ans: (b)

6. A point in a fluid flow where the flow has come to rest is called __________
   a) Pressure point
   b) Initial point
   c) Flow point
   d) Stagnation point

   Ans: (d)

7. Which of the following assumption is true about Bernoulli's equation..
   a) Flow is steady and irrotational
   b) Flow is incompressible and non viscous
   c) Flow is continuous and homogeneous with uniform velocity.
   d) All of the above

   Ans: (d)

8. The Bernoulli's is equation is based on....
   a) Conservation of mass
   b) Conservation of energy
   c) Conservation of both mass and energy
   d) None of the above

   Ans: (b)

9. Speed of the aeroplane is measured with the pitot tube.
   a) True
   b) False

   Ans: (a)

10. The principle of Venturimeter is based on the Bernoulli's equation.
    a) True
    b) False

    Ans: (a)

11. The pipe fitting with water having velocity 120 m/min and gauge pressure is 3.5 bar over the pipe is located at 8 m above the datum line. The total head of water in ....m.

a) 44.88m
b) 45.88m
c) 43.88m
d) 42.88m

Ans: (c)

12. The diverging angle of Venturimeter is less than converging angle.

a) True
b) False

Ans: (a)

13. The velocity of liquid at throat is higher than that of inlet in venturimeter.

a) True
b) False

Ans: (a)

14. The coefficient of discharge for orificemeter is about....

a) 0.95
b) 0.98
c) 0.85
d) 0.63

Ans: (d)

15. The coefficient of discharge is equal to...

a) Cd = cv/cc
b) Cd = cv – cc
c) Cd = cv + cc
d) Cd = cv × cc

Ans: (d)

16. The ratio of throat diameter and pipe diameter for Venturimeter is

a) 1:1
b) 1:2
c) 1:3
d) 2:1

Ans: (b)

17. The Venturimeter consists of which of the following parts.

a) Converging section
b) Throat section
c) Divergent section
d) All of the above

Ans: (d)

18. The average value of coefficient of discharge for Venturimeter is 0.98

a) True
b) False

Ans: (a)

19. The pitot tube having coefficient of friction is point 98 and used to measure velocity of water. The stagnation pressure recorded 3 m and static pressure is 2 m what is the velocity?

a) 4 m/s
b) 5 m/s
c) 4.34 m/s
d) 5.34 m/s

Ans: (c)

20. In case of venturimeter the pressure is maximum at midpoint of convergent section.

a) True
b) False

Ans: (a)

5.12.2: Theory Questions:

Q. 1 What do you understand by dynamics of fluid flow? How does it differ from the kinematics of fluid flow?

Q. 2 State the different forces considered in studying the equation of motion of a fluid. State the conditions under which each is significant.

Q. 3 Derive Euler's equation of motion for one dimensional flow along streamline. State clearly assumptions made and derive Bernoulli's equation.

Q. 4 Derive Euler's equation along a streamline in the form $\frac{dp}{\rho} + gdz + vdv = 0$.

Q 5 State Bernoulli's theorem.

Q 6 State Bernoulli's theorem. Explain significance of each term in Bernoulli's equation. State the assumptions made clearly.

Q. 7 Explain how Bernoulli's equation is applied to the real fluid flow problems.

Q. 8 Explain how Bernoulli's theorem, applied to two points in flow, is modified to account for

1. Loss of head
2. Installation of pump
3. Installation of a device like a turbine.
4. Non-uniform velocity variation in the pipe.

Q. 9. What is a venturimeter? Derive an expression for the discharge through venturimeter.

Q. 10 Explain the principle of venturimeter with a neat sketch.

Q. 11 Discuss the relative merits and demerits of venturimeter with orificemeter.

Q. 12 What is pitot tube? How will you determine the velocity at any point with the help of pitot – tube.

Q. 13 What is the difference between pitot tube and pitot static tube?

Q. 14 Define an orificemeter. Derive an expression for discharge through orificemeter.

Q. 15 what are Cd, Cc, Cv. Express the relation in them.

Q. 16 Draw a net sketch of venturimeter and orificemeter and explain their parts neatly.

5.12.2: Problems :

Q. 1 In an experiment on determination of hydraulic co-efficient of sharp-edged orifice, 2.5 cm of diameter it was found that the jet issuing horizontally under a head of 1 m travelled a horizontal distance of 1.6 m from vena contrata in a course of vertical drop of 0.7 m from the same point. If a flat plate is held nomal to jet at vena contrata, force 5.6 N would be exerted on the plate. Determine C_c, C_v, D_d for orifice.

Q. 2 In -a vertical pipe conveying oil of specific gravity 0.8, two pressure gauges have been installed at A and B, where diameters are 16 cm and 8 cm respectively. A is 2 m above B. The pressure gauge readings have shown that pressure at B is greater than at A by 0.981 N/cm^2. Neglecting all losses calculate flow rate.

Q. 3 Water is flowing through a tapered pipe having diameters 300 mm and 150 mm at sections 1 and 2 respectively. The discharge through the pipe is 40 litres/s. Section 1 is 10 m above datum and section 2 is 6 m above datum. Find the intensity of pressure at section 2 if that at section 1 is 400 kN/m^2. (May 2005, 6 Marks)

Q. 4 A venturimeter is used to measure the flow of petrol in a pipeline inclined at 35^0 to horizontal. The specific gravity of petrol is 0.81 and the throat area ratio is 4. If the difference in mercury levels in the gauge is 50 mm, calculate the flow in m^3/s if the pipe diameter is 300 mm. Take venturimeter constant as 0.95.

Q. 5 Find the rate of flow of water through a venturimeter fitted in a pipeline of diameter 30 cm. The ratio of diameter of throat and inlet to the venturimeter is $\frac{1}{3}$. The prossure at the iniet of the venturimeter is 1.4 bar (gauge) and vacuum at the throat is 37.5 cm of mercury. The coefficient of venturi is 0.98.

PRACTICAL: CALIBRATION OF VENTURIMETER

Objective: To determine the coefficient of venturimeter and calibrate it.

Apparatus: Venturimeter, differential manometer, measuring tank, stopwatch,

Theory: Calibration is a comparison between measurements – one of known magnitude or correctness made or set with one device and another measurement made in as similar a way as possible with a second device. Any instrument needs to be calibrated before its actual use.Venturimeter is a device used for measurement of discharge in a pipe line and works on the Principle of Bernoulli' theorem. The instrument consists of a short piece of a pipe which contracts up to section called as throat and then again enlarge up to the diameter as shown in figure. The conical portions joining the inlet and the throat and throat and outlet are called as converging cone and diverging cone respectively.

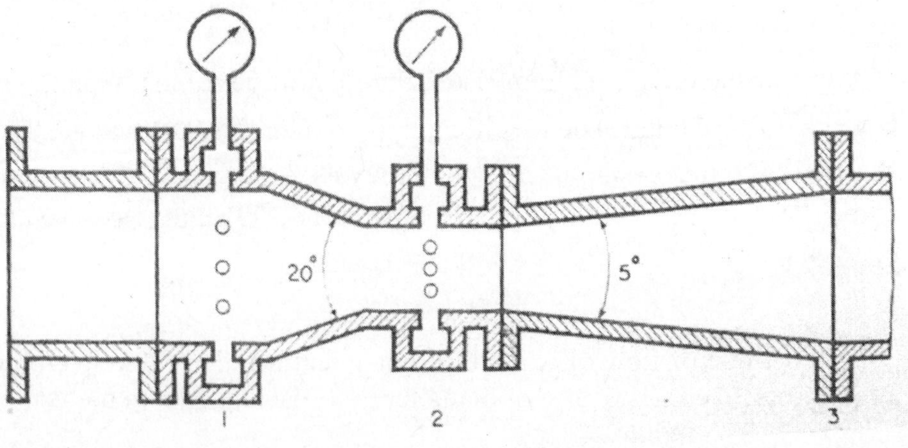

By contracting the passage of flow at the throat, the velocity of flow and hence the velocity head is increased. This increase in the velocity head causes change in the pressure head. The pressure difference thus created is measured generally by a U-tube manometer (differential) and the discharge through the pipe is calculated by the formula.

$$Q = KCH^{1/2}$$

Where,

Q = Discharge

K = Coefficient of discharge of venturimeter

H = difference of head in terms of water column between inlet and throat

C = Constant of venturimeter given by = $\frac{[a_1 \, a_2 (2g)^{0.5}]}{[a_1^2 - a_2^2]^{0.5}}$

Where,

a_1 = area of inlet which can found out from inlet diameter

$d_1 a_2$ = area of throat which can be found out throat diameterd_2

Actually the coefficient K is never unity and hence it is determined experimentally.

The above formula can be written a

Procedure:

1) Venturimeter is set up on the flow table and connected to the inlet pipe.
2) The manometer is then connected to the respective pressure tappings making sure that no airbubble is entrapped in the tube.
3) The water is allowed to flow through the venturimeter and the pressure difference is noted using the differential manometer.
4) The discharge is measured using measuring tank.
5) Time required to collect water in the measuring tank is noted
6) The procedure is repeated for different discharges

Observations:

1) Type of manometer : U Tube differential manometer
2) Inlet diameter of venturimeter = d_1= 0.029 m
3) Throat diameter of venturimeter = d_2 = 0.0145 m

Tabulated Calculations:

Sr. No.	Theoretical Discharge Q_{th} (m^3/s)	Actual Discharge Q_{act} (m^3 /s)	Coeff. of venturimeter K
1			
2			
3			
4			
5			
6			
7			
8			
9			
10			

Average of K =

Graph:

1. The graph between 'ln Qact' on y-axis and 'ln H' on x axis is plotted. The slope of graphgives value of index 'n'. While intercept gives the average value of 'M'. hence modified equation is $Q = MH^n$

$$\ln Q = \ln M + n \ln H$$

The above equation is similar to y = mx + c. Thus law of venturimeter or calibration equation is mentioned by taking antilog of ln M. The value of K is calculated by M = K C.

Calibration equation Q = ---------- H (*** Write value of M and n from graph)

Coefficient of discharge K = M/C

Intentionally Left Blank!

Table for graph calculations:

Sr. No.	ln Q_{act}	ln H	Sr. No.	ln Q_{act}	ln H

2. For calibration the graph between Qa on Y-axis and 'H' on x axis is plotted. The value of Qa is estimated by assuming certain value of H and compared with value of Qa obtained from laws of graph as established above. Both values must be same

Result:

Constant of venturimeter = C = Coefficient of venturimeter = K=

a) Avg. value from calculation =

b) Value from graph =

Conclusion:

1) Discharge formula = Q = ------- H(*** Write value of M and n from graph)

2) The calibration between actual discharge and differential head can be used for measurementof discharge whenever venturimeter is used.

Remark:

Coefficient of venturimeter normally lies between 0.91 to 0.99

PRACTICAL: CALIBRATION OF ORIFICEMETER

Objective: To determine the coefficient of orificemeter and calibrate it.

Apparatus: Orificemeter, differential manometer, measuring tank, stopwatch,

Theory: Calibration is a comparison between measurements – one of known magnitude or correctness made or set with one device and another measurement made in as similar a way as possible with a second device. Any instrument needs to be calibrated before its actual use. Orificemeter is a device used for measurement of discharge in a pipe line and works on the Principle of Bernoulli' theorem. The instrument consists of 'a' a thin plate with circular hole in it and held between flanges such that the orifice is concentric with the pipe.

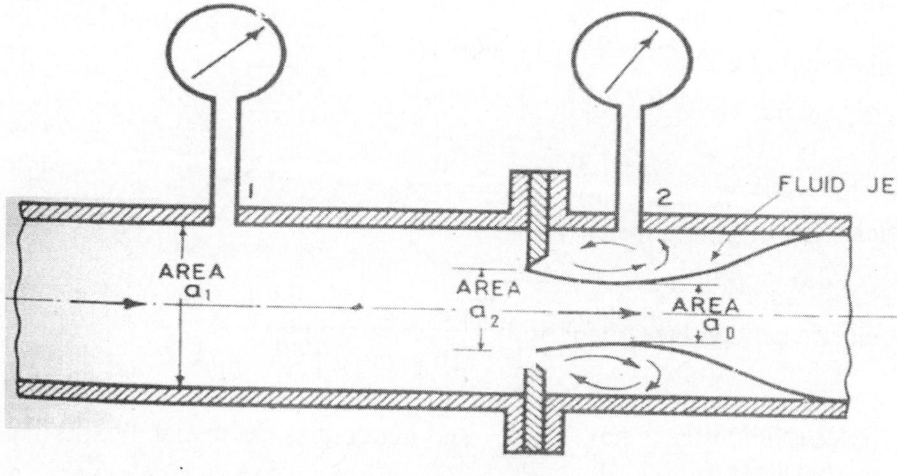

By contracting the passage of flow at the orifice, the velocity of flow and hence the velocity head is increased. This increase in the velocity head causes change in the pressure head. The pressure difference thus created is measured generally by a u-tube

$$Q = \frac{\left[C_d a_1 a_2 (2gH)^{1/2}\right]}{\left[a_1^2 - C_d^2 a_2^2\right]^{1/2}}$$

manometer (differential) and the discharge through the pipe is calculated by the formula.

Where C_d = coefficient of discharge,

a_1= area of inlet which can be found out from inlet diameter d_1,

a_2 = area of orifice which can be found out from orifice diameter d_2,

H = difference of head in terms of water column between inlet and orifice.

In the above equation put

$$K = C_d \frac{\left[a_1^2 - a_2^2\right]^{1/2}}{\left[a_1^2 - C_d^2 a_2^2\right]^{1/2}}$$

$$Q = KCH^{1/2}$$

Where,

Q = Discharge

K = coefficient of discharge of Orifice meter

C = constant of Orificemeter given by = $\frac{[a_1 \ a_2 \ (2g)^{0.5}]}{[a_1^2 - a_2^2]^{0.5}}$

Actually the coefficient K is never unity and hence it is determined experimentally. The above formula can be written as

$$Q = MH^n$$

Where M= K C

The constant M and n can be found out by plotting the result of the experiment on log-log scale. Once the M and n values are determined the orificemeter is said to be calibrated as discharge can then be easily calculated by measuring the head difference (as explained above) only.

Procedure:

1) Orificemeter is set up on the flow table and connected to the inlet pipe.
2) The manometer is then connected to the respective pressure tappings making sure that no air bubble is entrapped in the tube.
3) The water is allowed to flow through the orificemeter and the pressure difference is noted using the differential manometer.
4) The discharge is measured using measuring tank.
5) Time required to collect water in the measuring tank is noted
6) The procedure is repeated for different discharges

Observations:

1) Type of manometer : U Tube differential manometer
2) Inlet diameter = d_1= ------------ m
3) Diameter of orifice = d_2 = ------------ m

Observation table:

Sr. No.	**Head difference in terms of mercury in m (h)**	**Head difference in terms of water in m $H = h \times 12.6$**	**Discharge measurement**	
			Volume of water collected in m^3	**Time 't' in secs.**
1				
2				

3				
4				
5				

Sample calculations: For Observation No.

- Area of inlet $= a_1 = \frac{\pi}{4} d_1^2 =$ ---------------- m^2
- Area of orifice $= a_2 = \frac{\pi}{4} d_2^2 =$ ---------------- m^2
- Constant of orificemeter = C = $\frac{[a_1 \ a_2 \ (2g)^{0.5}]}{[a_1^2 - a_2^2]^{0.5}} =$
- Theoretical discharge $Q_{th} = CH^{0.5} =$
- Actual discharge Q_{act} = Volume / time
- Coefficient of orificemeter = K= Q_{act} / Q_{th}

Tabulated Calculations:

Sr. No.	Theoretical Discharge Q_{th} (m^3/s)	Actual Discharge Q_{act} (m^3/s)	Coeff. of Orificemeter K
1			
2			
3			
4			
5			

6			
7			

Average of K =

Graph:

1) The graph between 'ln Q_{act}' on y-axis and 'lnH 'on x axis is plotted. The slope of graph gives value of index 'n'. While intercept gives the average value of 'M'. hence modified equation is $Q = MH^n$

$\ln Q = \ln M + n \ln H$

The above equation is similar to $y = mx + c$.

Thus law of orificemeter or calibration equation is mentioned by taking antilog of ln M. The value of K is calculated by $M = K C$.

Calibration equation Q = ---------------- H (*** Write value of M and n from graph)

Coefficient of discharge $K = M/C$

Table for graph calculations:

Sr. No.	ln Q_{act}	ln H	Sr. No.	ln Q_{act}	ln H

2) For calibration the graph between Q_{act} on Y-axes and 'H'on x axis is plotted. The value of Q is estimated by assuming certain value of H and compared with value of Q_{act} obtained from laws of graph as established above. Both values must be same

Result:

Constant of orificemeter = C =

Coefficient of orificemeter = K =

a) Avg. value from calculation =
b) Value from graph =

Conclusion:

1) Discharge formula = Q = ------------ H (*** Write value of M and n from graph)
2) The calibration between actual discharge and differential head can be used for measurement of discharge whenever orifice meter is used

Remark: Coefficient of orifice meter normally lies between 0.61 to 0.65

Coefficient of orifice meter normally lies between 0.61 to 0.65

QR CODES FOR SUPPORTING VIDEO LINKS

6 DIMENSIONAL ANALYSIS

UNIT SPECIFICS

Through this unit we have discussed the following aspects:

- *Dimensional Analysis*
- *Non-dimensional numbers*

This is followed by large number of solved examples. The students are encouraged to solve the objective questions, long answer questions and numerical problems to judge ones understanding.

RATIONALE

This unit presents Buckingham's π theorem to perform dimensional analysis which is necessary especially when relationship between dependent and independent variables is to be established empirically. The six non-dimensional parameters which are important in establishing this relationship are then described.

PRE-REQUISITES

Mathematics: School level (Class VIII)

Physics: Dimensions (Class X)

UNIT OUTCOMES

List of outcomes of this unit is as follows:

(At the end of this unit, students will understand..)

U6-O1: Use of Buckingham's πtheorem for dimensional analysis

U6-O2: Non-dimensional parameters in fluid flow

Unit-6 Outcomes	EXPECTED MAPPING WITH COURSE OUTCOMES (1-WeakCorrelation; 2-Mediumcorrelation; 3-StrongCorrelation)					
	CO-1	CO-2	CO-3	CO-4	CO-5	CO-6
U6-O1	-	1	-	-	3	-
U6-O2	-	1	-	-	3	-

6.1 Introduction:

Dimensional analysis is a mathematical tool used to establish a relationship between various parameters governing a phenomenon with the help of their dimensions. Many fluid flow problems are very complex in nature making it very difficult to obtain their analytical solutions. The solutions of such problems are therefore based upon combination of physical analysis and experimental studies. The effect of different variables on the phenomenon under study and their interdependence is observed by experimentation. Dimensional analysis reduces the number of experiments by determining an empirical relation connecting the parameters and grouping them in various non-dimensional forms. Before the dimensional analysis is carried out, the variables controlling the phenomenon are identified and expressed in terms of primary dimensions. The phenomenon under consideration is then expressed by dimensionally homogeneous equations.

6.2 Dimensions:

As mentioned earlier, various physical quantities are required to describe a given phenomenon. These quantities can be described by a set of 'fundamental or primary units'. The primary units are mass, length, time and temperature represented by M, L, T, θ respectively. All other quantities like area, volume, velocity, force, acceleration etc. can be expressed in terms of these 'Fundamental quantities' and therefore termed as 'Derived quantities' or 'Secondary quantities'. The expression for the derived quantity in terms of fundamental quantities is called 'Dimension' of the derived quantity.

Thus, dimension of Velocity = $\left(\frac{\text{Distance}}{\text{Time}}\right)$ is expressed as $[LT^{-1}]$. Similarly, dimensions of acceleration (change in velocity/Time) and Force (mass × acceleration) can be expressed as $[LT^{-2}]$ and $[MLT^{-2}]$ respectively.

It may be noted that the dimensions of any physical quantity are independent of system of units i.e. whether the quantities are measured in SI, MKS, CGS or FPS systems. Table 6.3.1 presents the dimensions and units (SI system) of various physical quantities.

6.3 Dimensionally Homogeneous Equations:

It is imperative that for the method of dimensional analysis, the mathematical expression describing the phenomenon under study should be dimensionally homogeneous. i.e. dimensions of terms on left and right hand side of the equation are the same.

Consider the equation of flow over the weir.

$$Q = \frac{2}{3} C_d \,.\, \sqrt{2\,g} \,.\, L \,.\, H^{3/2} \qquad (6.1)$$

Where Q = Discharge $[LT^{-3}]$,

g = gravitational acceleration $[LT^{-2}]$,

L= Length [L], H = Head [L],

$\frac{2}{3}$, C_d, $\sqrt{2}$ is dimensionless quantity

Table 6.3.1: Dimensions of Physical quantities

Sr. No.	Quantity	Symbol	Dimensions MLT System	Unit of Measurement SI System
(A)	**Geometric**			
1	Length (any linear quantity)	l	L	m
2	Area	a, A	L^2	m^2
3	Volume	V	L^3	m^3
4	Slope	S	-	-
5	Angle	$\alpha\theta$	-	radians or degrees
(B)	**Kinematic**			
6	Time	T, t	T	sec.
7	Linear velocity	V, u	LT^{-1}	m/s
8	Angular velocity	ω	T^{-1}	rad/s
9	Linear acceleration	a	LT^{-2}	m/s^2
10	Angular acceleration	α	T^{-2}	rad/s^2
11	Gravitational acceleration	g	LT^{-2}	m/s^2
12	Discharge	Q	L^3T^{-1}	m^3/S
13	Discharge per unit width	q	L^2T^{-1}	m^2/s ($m^3/S/m$)
14	Kinematic viscosity	υ	L^2T^{-1}	m^2/s
(C)	**Dynamic**			
15	Mass	M, m	M	kg

16	Force	F	MLT^{-2}	N
17	Weight	W	MLT^{-2}	N
18	Mass density or specific mass	ρ	ML^{-3}	kg/m^3
19	Specific weight	γ	$ML^{-2}T^{-2}$	N/m^3
20	Specific gravity	S	-	-
21	Specific volume	∀ ∀	$M^{-1}L^3$ $M^{-1}L^2T^2$	m^3/kg m^3/N
22	Pressure	P	$ML^{-1}T^{-2}$	N/m^2 or Pascal (Pa)
23	Shear stress	τ	$ML^{-1}T^{-2}$	N/m^2
24	Dynamic viscosity	μ	$ML^{-1}T^{-1}$	$N\text{-}s/m^2$
25	Surface Tension	σ	MT^{-2}	N/m
26	Modulus of elasticity	K	$ML^{-1}T^{-2}$	N/m^2
27	Moment	M	ML^2T^{-2}	Nm
28	Momentum Impulse	M I	MLT^{-1}	kg m / s or N / s
29	Work, Energy	W, E	ML^2T^{-2}	Nm or Joule (J)
30	Torque	T	ML^2T^{-2}	Nm or Joule
31	Power	P	ML^2T^{-3}	N-m/s or watt (W)
32	Mass moment of inertia	I	ML^2	$kg - m^2$
33	Area moment of inertia	I	L^4	m^4

Note : Friction factor 'f', energy correction factor 'α', momentum correction factor 'β', efficiency 'η' are dimensionless quantities.

Substituting dimensions

Dimensions of L.H.S. = $[L^3 T^{-1}]$

Dimensions of R.H.S. = $[LT^{-2}]^{1/2} [L] . [L]^{3/2}$

$\therefore \quad [L^3 T^{-1}] = [LT^{-2}]^{1/2} [L] . [L]^{3/2}$

$\therefore \quad [L^3 T^{-1}] = [L^3 T^{-1}]$

This equation will hold good as long as all variables are consistent i.e. all variables should be substituted in same system of units. Dimensionally non-consistent parameters, for e.g. discharge Q in litre/s and g in m/s^2, if substituted the equation will not remain valid in the given form. This indicates that the equation of flow over the weir is dimensionally homogeneous.

However, there are several equations which are dimensionally non-homogeneous. For example, a steady uniform flow in an open channel with cement lining can be expressed (in MKS units) as,

$$U = 90\sqrt{RS} \qquad 6.2)$$

Where U = Velocity of flow $[LT^{-1}]$

R= Hydraulic Radius [L]

S= Channel slope (no dimension)

$\therefore \quad [LT^{-1}] \neq [L^{1/2}]$

This indicates that the above equation is dimensionally non-homogeneous. Consequently, it is implied that '90' has the dimensions of $\frac{[L^{1/2}]}{T}$ i.e. m$^{1/2}$/s in MKS units. Thus in FPS units, this equation will be expressed as $U = 163\sqrt{RS}$. ($\because$ 1 m = 3.281 ft)

6.4 Dimensional Analysis:

It is a systematic arrangement of variables, governing the phenomenon, with the help of dimensions. The dimensional analysis is useful in deriving equations in terms of non-dimensional parameters like Froude Number; Reynold's number, Mach Number etc. Due to grouping of variables to form dimensional parameters, number of variables are reduced making the experimentation more economical and less time consuming. One of the method used for dimensional analysis is Buckingham π Theorem.

6.4.1 Buckingham π Theorem:

If there are 'n' variables which govern a certain phenomenon and if these variables contain 'm' primary dimensions then the variable quantities can be expressed in terms of an equation containing 'n-m' dimensionless parameters.

The dimensionless groups are termed as π terms. Consider $X_1, X_2, \ldots X_n$ as variables involved in a problem. Therefore some functional relationships among them can be stated as $F(X_1, \ldots X_n) = 0$ or constant.

If these 'n' variables contain 'm' primary dimensions then according to Buckingham π theorem, there exists 'n-m' terms which are related by a functional relation,

$$f(\pi_1, \pi_2, \ldots \pi_{n-m}) = 0 \text{ or constant} \tag{6.3}$$

The π terms are formed by considering the 'm' quantities out of X quantities cumulatively containing all 'm' dimensions as repeating variables.

If X_1, X_2, X_3 be the repeating variables containing all 'm' dimensions collectively then π terms are formed as :

$$\begin{aligned} \pi_1 &= X_1^{x_1} \cdot X_2^{y_1} \cdot X_3^{z_1} \cdot X_4 \\ \pi_2 &= X_1^{x_2} \cdot X_2^{y_2} \cdot X_3^{z_2} \cdot X_5 \\ &\vdots \\ \pi_{n-m} &= X_1^{x_{n-m}} \cdot X_2^{y_{n-m}} \cdot X_3^{z_{n-m}} \cdot X_{n-m} \end{aligned} \tag{6.4}$$

Where $X_4, X_5, \ldots X_{n-m}$ are non-repeating variables involved in a problem. After substituting dimensions of all X quantities, equations are formed by equating exponents of π terms (M, L, T) to zero. The solutions of these equations yield exponents X_1, X_2, X_3 etc.

6.4.2 Steps for Application of Buckingham π Theorem:

1. Make a list of all 'n' variables involved in the given problem and write the functional relationship.
2. Identify the dependent variables (Generally the variable on L.H.S. of the equation asked to derive).
3. Write down dimensions of all 'n' variables and find out number of m primary dimensions involved. Number of π terms = n - m
4. Choose 'm' number of repeating variables (i.e. number of repeating variables = number of primary dimensions involved in the given problem).

Guidelines:

1. None of the repeating variables should be dimensionless.
2. No two of the repeating variables should have the same dimensions i.e. they should not form a dimensionless parameter among themselves.
3. Repeating variables should contain all 'm' dimensions collectively.
4. Dependent variables should not be selected as repeating variables.
5. Generally, the first repeating variable should be from those variables which describe 'Geometry of flow, e.g. length, diameter, height, breadth etc.
6. Second repeating variable should be from those variables which describe 'flow property' i.e. velocity, acceleration, force, power etc.
7. The third variable should be from those variables which describe 'fluid property', e.g. viscosity, mass density, surface tension etc.

   Note: Generally, mass density ρ, velocity v and length l are taken as repeating variables for fluid mechanics problems. Generally, the first term on R.H.S, of the given problem indicates repeating variables e.g. prove that $R = \rho L^2 v^2 \phi (R_e, M)$. Here ρ, v, L are the repeating variables)

8. Raise the repeating variables to unknown powers and combine them with other variables one after the other to form dimensionless groups.

9 Equate exponents of π term (M, L, T) to zero. Express the final expression in terms of all 'n–m' dimensionless groups operations with π term:

i) A dimensionless variable is a π term itself. e.g. θ^0, efficiency η.

ii) Ratio of two variables with same dimensions forms a π term by observation e.g. L/D, H/D

iii) Any π term can be multiplied or divided by any numerical number or by any other π term $3\pi_1$, $\pi_2/5$, π_1/π_2, π_3/π_4.

iv) A π term can be raised to any power. e.g. $\pi_1^2, \pi_1^{-2}, \pi_1^{3/2}$.

The above steps can be understood better by solving a numerical problem as given below.

Ex. 1. Drag on a body depends upon its characteristic length *l*, speed v, mass density ρ, viscosity μ and gravitational acceleration g. Obtain an expression for drag F in terms of dimensionless parameters using Buckingham's π theorem.

Solution :

$$f(\text{F}, l, \text{v}, \rho, \mu, \text{g}) = 0 \text{ or constant}$$

Number of variables = 6

Dimensions

F	l	v	ρ	μ	g
$[MLT^{-2}]$	$[L]$	$[LT^{-1}]$	$[ML^{-3}]$	$[ML^{-1}T^{-1}]$	$[LT^{-2}]$

Number of ' m ' primary dimensions involved = 3

∴ Number of π terms = 6 – 3 = 3

∴ $f_1(\pi_1, \pi_2, \pi_3) = 0$ or constant.

Selecting ρ, v, l as repeating variables,

$$\pi_1 = \rho^{x_1} v^{y_1}\, l^{z_1}\, \text{F}$$

$$[M^0 L^0 T^0] = [ML^{-3}]^{x_1}[LT^{-1}]^{y_1}[\,L]^{z_1}[ML\,T^{-2}]$$

(equating powers of M, L and T respectively)

$$0 = x_1 + 1, \qquad 0 = -3x_1 + y_1 + z_1 + 1, \qquad 0 = -y_1 - 2$$

$$\therefore \quad x_1 = -1, \qquad y_1 = -2, \qquad z_1 = -2$$

$$\therefore \quad \pi_1 = \rho^{-1} v^{-2} l^{-2} F = \frac{F}{\rho v^2 \rho^2}$$

$$\pi_2 = \rho^{x_2} v^{y_2} l^{z_2} \mu$$

$$[M^0 L^0 T^0] = [ML^{-3}]^{x_2} [LT^{-1}]^{y_2} [L]^{z_2} [ML^{-1} T^{-1}]$$

$$\therefore \quad 0 = x_2 + 1, \qquad 0 = -3x_2 + y_2 + z_2 - 1 \qquad 0 = -y_2 - 1$$

$$x_2 = -1 \qquad z_2 = -1 \qquad y_2 = -1$$

$$\therefore \quad \pi_2 = \rho^{-1} v^{-1} l^{-1} \mu = \frac{\mu}{\rho v l} = \frac{1}{R_e} = R_e \left(\pi_2 = \frac{1}{\pi_2}\right)$$

$$\therefore \quad \pi_3 = \rho^{x_3} v^{y_3} l^{z_3} g$$

$$[M^0 L^0 T^0] = [ML^{-3}]^{x_3} [L\,T^{-1}]^{y_3} [L^{z_3} [LT^{-2}]$$

$$\therefore \quad 0 = x_3, \qquad 0 = -3x_3 + y_3 + z_3 + 1 \qquad 0 = -y_3 - 2$$

$$\therefore \quad x_3 = 0, \qquad z_3 = 1, \qquad y_3 = -2$$

$$\therefore \quad \pi_3 = \rho^0 v^{-2} l^L g = \frac{gl}{v^2} = \frac{v}{\sqrt{gl}} = Fr \ldots \left(\pi_3 = \frac{1}{\pi_3}\right)$$

$$\therefore \quad f_1\left(\frac{F}{\rho v^2 J^2}, R_e, F_r\right) = 0 \text{ or constant}$$

$$\therefore \quad \frac{F}{\rho f^2 v^2} = \phi\ (R_e, F_r)$$

$$\therefore \quad F = \rho l^2 v^2 \phi\ (R_e, F_r)$$

Thus, Buckingham π theorem yields the value of exponents explicitly. Also the possible number of π terms involved is known before hand.

6.5 Dimensionless Parameters:

The dimensionless parameters required for analysis of fluid flow are generally ratios of forces per unit volume. These parameters or merely 'number' are,

1 Froude Number, $F_r = \left(\frac{\text{inertia force}}{\text{gravity force}}\right)^{1/2}$

2 Reynold's Number, $R_e = \left(\frac{\text{inertia force}}{\text{viscous force}}\right)$

3 Mach Number, $M = \left(\frac{\text{inertia force}}{\text{elastic force}}\right)^{1/2}$

4 Weber Number, $W = \left(\frac{\text{inertia force}}{\text{surface tension force}}\right)^{1/2}$

5 Euler's Number, $E = \left(\frac{\text{inertia force}}{\text{pressure force}}\right)^{1/2}$

These five numbers are explained briefly below.

6.5.1 Froude Number F_r :

$$F_r = \left(\frac{\text{inertia force}}{\text{gravity force}}\right)^{1/2} = \left(\frac{\rho L^2 v^2}{\rho L^3 g}\right)^{1/2}$$

$$\therefore \quad F_r = \frac{v}{\sqrt{gL}} \tag{6.5}$$

(Note : 'L' can be characteristic dimension like length, diameter, hydraulic depth etc.)

Application: Free surface flows which are dominated by gravity e.g. open channel flow.

6.5.2 Reynold's Number R_e:

$$R_e = \left(\frac{\text{inertia force}}{\text{viscous force}}\right) = \left(\frac{\rho L^2 v^2}{\mu v L}\right) = \frac{\rho v}{\mu} = \frac{v L}{\nu} \tag{6.6}$$

Application: Incompressible fluid flows with dominant viscous force e.g. pipe flow, motion of fully submerged bodies like submarine, Torpedo, airplane etc.

6.5.3 Mach Number M :

$$M = \left(\frac{\text{inertia force}}{\text{elastic force}}\right)^{1/2} = \left(\frac{\rho L^2 v^2}{k L^2}\right)^{1/2} = \frac{v}{\sqrt{k/\rho}} \tag{6.7}$$

Application: Compressible flows, water hammer phenomenon.

6.5.4 Weber Number W :

$$W = \left(\frac{\text{inertia force}}{\text{surface tension force}}\right)^{½}$$

$$= \left(\frac{\rho L^2 v^2}{\sigma L}\right)^{½} = \frac{v}{\sqrt{\frac{\sigma}{\rho L}}} \tag{6.8}$$

Application: Study of droplets or capillary rise where a thin film is involved or depth of flow is small.

6.5.5 Euler Number E :

$$E = \left(\frac{\text{inertia force}}{\text{pressure force}}\right)^{½} = \left(\frac{\rho L^2 v^2}{p L^2}\right)^{½} = \frac{v}{\sqrt{p/\rho}} \tag{6.9}$$

Application: Study of cavitation phenomenon where pressure force is dominant.

6.6 Solved Problems :

Ex. 6.1: The variables controlling the motion of a floating vessels through water are the of the water and the gravitational acceleration g. Derive an expression for F by dimensional analysis.

Solution:

$f(\text{F, V}, l, \rho, \mu, \text{g}) = 0$ or constant

Dimensions

F	V	l	ρ	μ	g
[MLT^{-2}]	[LT^{-1}]	[L]	[ML^{-3}]	[$ML^{-1}T^{-1}$]	[LT^{-2}]

Number of variables n = 6

Number of primary dimensions m = 3

∴ Number of π terms n – m = 3

∴ $f(\pi_1, \pi_2, \pi_3) = 0$ or constant

Take ρ, V, l as repeating variables.

$$\pi_1 = [\rho^{x_1} V^{y_1}]^{z_1}\ F$$

$$[M^0 L^0 T^0] = [ML^{-3}]^{x_1} [LT^{-1}]^{y_1} [\ L^{z_1} [MLT^{-2}]$$

$$x_1 + 1 = 0, \quad -3\,x_1 + y_1 + z_1 + 1 = 0, \quad -y_1 - 2 = 0$$

∴ $x_1 = -1, \quad z_1 = -2, \quad y_1 = -2$

∴ $\pi_1 = \rho^{-1} V^{-2}\ J^{-2} F^{-2} = \dfrac{F}{\rho V^2 \rho^2}$

Now, $\pi_2 = \rho^{x_2} V^{y_2} l^{z_2} \mu$

$$[M^0\, L^0\, T^0] = [ML^{-3}]^{x_2} [LT^{-1}]^{y_2} [L]^{z_2} [ML^{-1}\ T^{-1}]$$

$$x_2 + 1 = 0, \quad -3\,x_2 + y_2 + z_2 - 1 = 0, \quad -y_2 - 1 = 0$$

	∴ $x_2 = -1, \quad z_2 = -1, \quad y_2 = -1$ ∴ $\pi_2 = \rho^{-1} V^{-1} l^{-1} \mu = \frac{\mu}{\rho V l}$ Now, $\pi_2 = \rho^{x_1} v^{y_3 T_3} g$ $[M^0 L^0 T^0] = [ML^{-3}]^{x_3} [LT^{-1}]_3 [L]^{\alpha_3} [LT^{-2}]$ $x_3 = 0, \quad -3x_3 + y_3 + z_3 + 1 = 0, \quad -y_3 - 2 = 0$ ∴ $x_3 = 0, \quad z_3 = 1, \quad y_3 = -2$ ∴ $\pi_3 = \rho^0 V^{-2} l^1 g = \frac{gl}{V^2}$ ∴ $f_1\left(\frac{F}{\rho V^2 l^2}, \frac{\mu}{\rho V l}, \frac{gl}{V^2}\right) = 0$ or constant ∴ $F = \rho V^2 \rho \phi \left[\frac{\mu}{\rho V l}, \frac{gl}{V^2}\right]$

Ex. 6.2:	The thrust F of supersonic air-craft during light is dependent on the length of aircraft (L), air-craft velocity (V), air viscosity (μ), air density (ρ) and bulk modulus of air (K). Using Buckingham π theorem, show that the rational equation for thrust is given by : $$F = \rho L^2 V^2 \phi \left[R_e, \frac{K}{\rho V^2}\right]$$ **Solution:** $f(F, L, V, \mu, \rho, K) = 0$ or constant Dimensions F L V μ ρ K

$$[LT^{-2}] \; [L] \; [LT^{-1}] \; [ML^{-1}T^{-1}] \; [ML] \; [ML^{-1}T^{-2}]$$

Number of variables n = 6

Number of primary dimensions m = 3

∴ Number of π terms = 3

∴ $f(\pi_1, \pi_2, \pi_3) = 0$ or constant

Take ρ, V, l as repeating variables.

$$\pi_1 = \rho^{x_1} V^{y_1} L^{z_1} F$$

$$[M^0 L^0 T^0] = [ML^{-3}]^{x_1} [LT^{-1}]^{y_1} [L]^{z_1} [MLT^{-2}]$$

$$x_1 + 1 = 0, \quad -3x_1 + y_1 + z_1 + 1 = 0, \quad -y_1 - 2 = 0$$

∴ $x_1 = -1, \quad z_1 = -2, \quad y_1 = -2$

∴ $$\pi_1 = \rho^{-1} V^{-2} L^{-2} F = \frac{F}{\rho V^2 L^2}$$

Now, $$\pi_2 = \rho^{x_2} V^{y_2} L^{z_2} \mu$$

$$[M^0 L^0 T^0] = [ML^{-3}]^{x_2} [LT^{-1}]^{y_2} [L^{z_2} [ML^{-1} T^{-1}]$$

$$x_2 + 1 = 0, \quad -3x_2 + y_2 + z_2 - 1 = 0, \quad -y_2 - 1 = 0$$

∴ $x_2 = -1, \quad z_2 = -1, \quad y_2 = -2$

∴ $$\pi_2 = \rho^{-1} V^{-1} L^{-1} \mu = \frac{\mu}{\rho V L}$$

$$\frac{1}{\pi_2} = \frac{\rho V L}{\mu} = R_e$$

Now, $$\pi_3 = \rho^{x_3} V^{y_3} L^{z_3} \; K$$

$$[M^0 L^0 T^0] = [MLL^{-3}]^{x_3} [LT^{-1}]^{y_3} [L]^{x_3} [ML^{-1} T^{-2}]$$

$$x_3 + 1 = 0, \quad -3x_3 + y_3 + z_3 - 1 = 0, \quad -y_3 - 2 = 0$$

	$\therefore \quad x_3 = -1, \quad z_3 = 0, \quad y_3 = -2$ $\therefore \quad \pi_3 = \rho^{-1} V^{-2} L^0 K = \frac{K}{\rho V^2}$ $\therefore \quad \frac{F}{\rho V^2 L^2} = \phi\left(R_e, \frac{K}{\rho V^2}\right)$ $\therefore \quad F = \rho V^2 L^2 \phi\left(R_e, \frac{K}{\rho V^2}\right)$

Ex. 6.3:	A viscous fluid is confined between two long cylinders. A torque per unit length 'J' is required to turn the inner cylinder at constant angular velocity 'ω'. The cylinder radii are 'r' and 'R' and the fluid 'viscosity is 'μ'. Set up a nondimensional equation for the set of parameters given. If both radii are doubled, how does it affect T. If the viscosity is halved, what effect does it have on T. **Solution :** $f(J, \omega, r, R, \mu) = 0$ or constant Dimensions $J \quad \omega \quad r \quad R \quad \mu$ $[MLT^{-2}] \; [L] \; [t] \; [L] \; [ML^{-1}T^{-1}]$ Number of variables n = 5 Number of primary variables m = 3 $\therefore$ Number of π terms = 2 $f(\pi_1, \pi_2) = 0$ or constant Taking ω, r, μ as repeating variables. $\pi_1 = \omega^{x_1} r^{y_1} \mu^{z_1} J$ $[M^0 L^0 T^0] = [T^{-1}]^{x_1} [L]^{y_1} [ML^{-1} T^{-1}]^{z_1} [MLT^{-2}]$

	$-x_1 - z_1 - 2 = 0, \quad y_1 - 1 + 1 = 0, \quad - z_1 - 2 = 0$
	$\therefore \quad x_1 = 0, \quad y_1 = 0, \quad z_1 = -2$
	$\therefore \quad \pi_1 = \omega^0\, r^0\, \mu^{-2}\, J = \frac{J}{\mu^2}$
	Now $\quad \pi_2 = \frac{R}{r}$ by observation
	$\therefore \quad f\left(\frac{J}{\mu^2}, \frac{R}{r}\right) = 0$ or constant
	$\therefore \quad J = \mu^2\phi\left(\frac{R}{r}\right)$
	i) If both radii are doubled torque will not change.
	ii) If μ is halved, J will be reduced by $\frac{1}{4}$.

Ex. 6.4 :	The pressure drop per unit length ($\Delta p/L$) for an incompressible fluid flow through a pipe depends upon the pipe diameter (D), the pipe roughness (E), velocity of the fluid (V), fluid density (ζ) and its viscosity (μ), Using the Buckingham π - theorem show that the relation between the variables is given by: $$\frac{Ap}{L} = \frac{\zeta V^2}{d} f\left(\frac{E}{D} \cdot \frac{\mu}{\zeta VD}\right)$$ **Solution :** $f(\frac{\Delta p}{L}$, D, $\in$, V, ζ, μ) = 0 or constant Dimensions

$$\frac{\Delta p}{L} \quad D \quad \in \quad V \quad \zeta \quad \mu$$

$$[ML^{-2}T^{-2}] \quad [L] \quad [L] \quad [LT^{-1}] \quad [ML^{-3}] \quad [ML^{-1}T^{-1}]$$

Number of variables n = 6

Number of primary dimensions m = 3

$\therefore$ Number of π terms = 3

$\therefore$ $f(\pi_1, \pi_2, \pi_3) = 0$ or constant

Taking ρ, V, D as repeating variables.

$$\pi_1 = \zeta^{x_1} V^{y_1} D^{z_1} \frac{\Delta P}{L}$$

$$[M^0 L^0 T^0] = [ML^{-3}]^{x_1} [LT^{-1}]^{y_1} [L]^{z_1} [ML^{-2} T^{-2}]$$

$$x_1 + 1 = 0, \quad -3x_1 + y_1 + z_1 - 2 = 0, \quad -y_1 - 2 = 0$$

$\therefore$ $x_1 = -1, \quad z_1 = 1, \quad y_1 = -2$

$\therefore$ $$\pi_1 = \zeta^{-1} V^{-2} D^1 \frac{\Delta P}{L} = \frac{(\Delta P/L) \cdot D}{\zeta V^2}$$

Now, $\pi_2 = \dfrac{\varepsilon}{D}$ by observation.

Now, $\pi_2 = \zeta^{x_2} v^{y_2} D^{z_2} \mu$

$$[M^0 L^0 T^0] = [ML^{-3}]^{x_2} [LT^{-1}]^{\frac{1}{2}} [L]^{2\ 2} [ML^{-1} T^{-1}]$$

$$x_2 + 1 = 0, \quad -3x_2 + y_2 + z_2 - 1 = 0, \quad -y_2 - 1 = 0$$

$\therefore$ $x_2 = -1, \quad z_2 = -1, \quad y_2 = -1$

$\therefore$ $$\pi_3 = \zeta^{-1} V^{-1} D^{-1} \mu = \frac{\mu}{\zeta VD}$$

$$\therefore f_1\left(\frac{(\Delta P/L) \cdot D}{\zeta V^2}, \frac{E}{D}, \frac{\mu}{\zeta VD}\right) = 0 \text{ or constant}$$

	$\therefore \quad \frac{\Delta P}{L} = \frac{\zeta V^2}{D}\phi\left(\frac{E}{D}, \frac{\mu}{\zeta VD}\right)$

Ex. 6.5:	Using the Buckingham π theorem show that the velocity U through a circular orifice is given by, $u = \sqrt{gH}\,\phi\left[\frac{\alpha}{H}, \frac{\mu}{\rho VH}\right]$ where H is the head of the orifice, α diameter of orifice, p density and μ viscosity. **Solution :** $f(u, g, H, \alpha, \mu, \rho) = 0$ or constant Dimensions u g H α μ ρ $[LT^{-1}]$ $[LT^{-2}]$ $[L]$ $[L]$ $[ML^{-1}T^{-1}]$ $[ML^{-3}]$ Number of variables n = 6 Number of primary dimensions m = 3 $\therefore$ Number of π terms = 3 $\therefore$ $f(\pi_1, \pi_2, \pi_3) = 0$ or constant Taking ρ, u, H as repeating variables. $\pi_1 = \rho^{x_1} u^{y_2} H^{z} g$ $[M^0 L^0 T^0] = [ML^{-3}]^{x_1}[LT^{-1}]^{y_2}[L]^{z_1}[LT^{-2}]$ $x_1 = 0, \quad -3x_1 + y_2 + z_1 + 1 = 0, \quad -y_2 - 2 = 0$ $\therefore \quad x_1 = 0, \quad z_1 = 1, \quad y_2 = -2$

	$\therefore \quad \pi_1 = \rho^0 u^{-2} H^1 g = \frac{gH}{u^2} \quad \therefore \sqrt{\pi_1} = \frac{\sqrt{gH}}{u}$
	Now, $\quad \pi_2 = \frac{\alpha}{H}$ by observation.
	Now, $\quad \pi_2 = \rho^{x_2} \cdot u^{y_2} H^2 \mu$
	$[M^0 L^0 T^0] = [ML^{-3}]^{x_2} [LT^{-1}]^{y_2} [L]^{x_2} [M^1 L^{-1} T^{-1}]$
	$x_2 + 1 = 0, \quad -3x_2 + y_2 + z_2 - 1 = 0, \quad -y_2 - 1 = 0$
	$\therefore \quad x_2 = -1, \quad z_2 = -1, \quad y_2 = -1$
	$\therefore \quad \pi_3 = \rho^{-1} u^{-1} H^{-1} \mu = \frac{\mu}{\rho u H}$
	$\therefore \quad f_1\left(\frac{\sqrt{gH}}{u}, \frac{\alpha}{H}, \frac{\mu}{\rho VH}\right) = 0$ or constant
	$\therefore \quad \sqrt{gH} = \phi\left[\frac{\alpha}{H}, \frac{\mu}{\rho VH}\right]$
	$\therefore \quad u = \sqrt{gH}\,\phi\left[\frac{\alpha}{H}, \frac{\mu}{\rho VH}\right]$

Ex. 6.6 :	The resisting torque T of a lubricated joumal bearing depends on the joumal diameter D, clearance c, length l, speed of rotation N, viscosity of oil μ and load W show that, $$\frac{T}{WD} = \phi\left(\frac{l}{D}, \frac{C}{D}, \frac{\mu D^2 N}{W}\right)$$ **Solution :** $f(T, D, C, l, N, \mu, w) = 0$ or constant

Dimensions

T	D	C	l	N	μ	w
$[ML^2T^{-2}]$	[L]	[L]	[L]	$[T^{-1}]$	$[ML^{-1}T^{-1}]$	$[MLT^{-2}]$

Number of variables n = 7.

Number of primary dimensions m = 3.

Number of π tems n – m = 4.

$$\therefore \quad f(\pi_1, \pi_2, \pi_3, \pi_4) = 0 \text{ or constant}$$

Take W, N, D as repeating variables.

$$\pi_1 = W^{x_1} N^{y_1} D^{z_1} T$$

$$[M^0 L^0 T^0] = [MLTT^{-2}]^{x_1}[T^{-1}]^{y_1}[L]^{z_1}[ML^2 T^{-2}]$$

$$0 = x_1 + 1, \quad 0 = x_1 + z_1 + 2, \quad 0 = -2x_1 - y_1 - 2$$

$$\therefore \quad x_1 = -1, \quad z_1 = -1, \quad y_1 = 0$$

$$\therefore \quad \pi_1 = W^{-1} N^0 D^{-1} T = \frac{T}{WD}$$

$$\therefore \quad \pi_2 = \frac{l}{D'}, \quad \pi_3 = \frac{C}{D}$$

$$\therefore \quad \pi_4 = W^{x_2} N^{y_2} D^2 \mu$$

$$[M^0 L^0 T^0] = [MLT^{-2}]^{x_1}[T^{-1}]^{y_1}[L]^{z_1}[ML^{-1} T^{-1}]$$

$$0 = x_1 + 1, \quad 0 = x_1 + z_1 - 1, \quad 0 = -2x_1 - y_1 - 1$$

$$\therefore \quad x_1 = -1, \quad z_1 = 2 \quad y_1 = 1$$

$$\therefore \quad f\left(\frac{T}{WD}, \frac{l}{D}, \frac{C}{D}, \frac{\mu ND^2}{W}\right) = 0 \text{ or constant}$$

$$\frac{T}{WD} = \phi\left(\frac{l}{D}, \frac{C}{D}, \frac{\mu ND^2}{W}\right)$$

Ex. 6.7:	Power P, developed by a water turbine depends upon rotation speed N, operating head H, diameter D, breadth 'B' of runner, density p, viscosity μ and gravity g. Show $$p = \rho D^3 N^3 \phi \left[\frac{N}{D}, \frac{B}{D}, \frac{\rho D^2 N}{\mu}, \frac{ND}{\sqrt{gN}}\right]$$ **Solution :** f(P, N, H, D, B, ρ, μ, g) = 0 or constant Dimensions P N H D B ρ μ g $[ML^2T^{-3}]$ $[T^{-1}]$ [L] [L] [L] $[ML^{-3}]$ $[MLT^{-1}T^{-1}]$ $[LT^{-2}]$ Number of variables n = 8. Number of primary dimensions m = 3 ∴ Number of π terms = 5 ∴ $f(\pi_1, \pi_2, \pi_3, \pi_4, \pi_5, \pi_6, \pi_7, \pi_8) = 0$ or constant Take ρ, N, D as repeating variable. $$\pi_1 = \rho^{x_1} N^{y_1} D^{z_1} P$$ $$[M^0 L^0 T^0] = [ML^{-3}]^{x_1} [T^{-1}]^{y_1} [L]^{z_1} [ML^2 T^{-3}]$$ $$0 = x_1 + 1, \quad 0 = -3x_1 + z_1 + 2, \quad 0 = -y_1 - 3$$ ∴ $x_1 = -1, \quad z_1 = -5, \quad y_1 = -3,$ ∴ $\pi_1 = \rho^{-1} N^{-5} D^{-5} P$

∴ $$\pi_1 = \frac{P}{\rho N^5 D^5}$$

Now $$\pi_2 = \frac{H}{D}, \ \pi_3 = \frac{B}{D} \text{ by observation}$$

$$\pi_4 = \rho^{x_2} N^{y_2} D^{z_2} \mu$$

∴ $$[M^0 L^0 T^0] = [M^{-3}]^{x_2} [T^{-1}]^{y_2} [L]^{z_2} [MLL^{-1}L^{-1}]$$

$$0 = x_2 + 1, \quad 0 = -3x_2 + z_2 - 1, \quad 0 = -y_2 - 1$$

∴ $$x_2 = -1, \qquad z_2 = -2, \qquad y_2 = -1$$

∴ $$\pi_4 = \rho^{-1} N^{-1} D^{-2} \mu = \frac{\mu}{\rho N D^2}$$

Now, $$\pi_5 = \rho^{x_3} N^{y_3} D^{z_3} g$$

$$[M^0 L^0 T^0] = [M^{-3}]^{x_3} [T^{-1}]^{y_3} [L]^{z_3} [LT^{-2}]$$

$$0 = x_3, \quad 0 = -3x_3 + z_3 + 1, \quad 0 = -y_3 - 2$$

∴ $$z_3 = -1, \qquad y_3 = -2$$

∴ $$\pi_5 = \rho^0 N^{-2} D^{-1} g = \frac{g}{N^2 D}$$

$$\therefore f\left(\frac{p}{\rho N^5 D^5}, \frac{H}{D}, \frac{B}{D}, \frac{\mu}{\rho N D^2}, \frac{g}{N^2 D}\right) = 0 \text{ or constant}$$

∴ $$P = \rho N^3 D^5 \phi\left(\frac{H}{D}, \frac{B}{D}, \frac{8}{N^2 D}, \frac{\mu}{\rho N D^2}\right)$$

Now $$\pi_2 \times \pi_5 = \frac{H}{D} \times \frac{g}{N^2 D} = \frac{gH}{N^2 D^2}$$

∴ $$\frac{1}{\pi_1 \pi_5} = \frac{N^2 D^2}{gH}$$

	$(\pi_1 \times \pi_5)^{1/2} = \frac{ND}{\sqrt{gH}}$ $\therefore \quad P = \rho N^3 D^5 \phi\left(\frac{H}{D}, \frac{B}{D}, \frac{\rho N D^2}{\mu}, \frac{ND}{\sqrt{gH}}\right)$

Ex. 6.8 :	A supersonic plane of length L moves with speed V through air of density ρ, method show that dimensionless groups obtained are Mach number, Reynold's number and Drag coefficient $\frac{F_D}{\rho L^2 V^2}$ **Solution :** $f(L, V, \rho, \mu, k, F_D) = 0$ or constant Dimensions L: [L], V: $[LT^{-1}]$, ρ: $[ML^{-3}]$, μ: $[ML^{-1}T^{-1}]$, k: $[ML^{-1}T^{-2}]$, F_D: $[MLT^{-2}]$ Number of variables n = 6 Number of primary dimensions m = 3 $\therefore$ Number of π terms = 3 Taking ρ, V, L as repeating variables. $\pi_1 = \rho^{x_1} N^{y_1} L^2 \mu$ $[M^0 L^0 T^0] = [ML^{-3}]^{x_1} [LT^{-1}]^{y_1} [L]^{z_1} [ML^{-1} T^{-1}]$ $x_1 + 1 = 0, \quad -3x_1 + y_1 + z_1 - 1 = 0, \quad -y_1 - 1 = 0$ $\therefore \quad x_1 = -1, \quad 3 - 1 + z_1 - 1 = 0, \quad y_1 = -1$ $\therefore \quad z_1 = -1$

	$\therefore \quad \pi_1 = \rho^{-1} V^{-1} L^{-1} \mu = \frac{\mu}{\rho VL}$ $= \frac{1}{R_e}$ (First dimensionless group is Reynold's number) $\therefore \quad \frac{1}{\pi_1} = R_e$ $\pi_2 = \rho^{x_2} V^{y_2} L^z k$ $[M^0 L^0 T^0] = [ML^{-3}]^{x_2} [LT^{-1}]^{y_2} [L]^2 [ML^{-1} T^{-2}]$ $x_2 + 1 = 0, \quad -3x_2 + y_2 + z_2 - 1 = 0, \quad -y_2 - 2 = 0$ $\therefore \quad x_2 = -1, \quad 3 - 2 + z_2 - 1 = 0, \quad y_2 = -2,$ $\therefore \quad z_2 = -1$ $\therefore \quad \pi_2 = \rho^{-1} V^{-2} L^0 k = \frac{k}{\rho V^2}$ $\frac{1}{\pi^2} = \frac{\rho V^2}{k} = \frac{V^2}{k/\rho}$

Ex. 6.9:	Discharge Q of a centrifugal pump can be assumed to be dependent on speed N rpm. Using t, method show that, $Q = ND^3 \phi \left(\frac{gH}{N^2 D^2}, \frac{\gamma}{N\, N^2} \right)$ **Solution :** $f(Q, \rho, \mu, P, D, N) = 0$ or constant Dimensions Q ρ μ P D N $[L^3T^{-1}]$ $[ML^{-3}]$ $[ML^{-1}T^{-1}]$ $[ML^{-1}T^{-2}]$ $[L]$ $[T^{-1}]$

	Number of variables n = 6
	Number of primary dimensions m = 3
	Number of terms = 3
	$\therefore \quad f(\pi_{1,}\pi_{2,}\pi_3) = 0$ or constant
	Taking ρ, N, D as repeating variables.
	$\pi_1 = \rho^{x_1} N^{y_1} D^{z_1} Q$
	$[M^0 L^0 T^0] = [M^{-3}]^{x_1}[T^{-1}]^{y_1}[L]^2[L^3T^{-1}]$
	$x_1 = 0, \quad -3x_1 + z_1 + 3 = 0, \quad -y_1 - 1 = 0$
	$\therefore \quad z_1 = -3 \qquad y_1 = -1$
	$\therefore \quad \pi_1 = \rho^0 N^{-1} D^{-3} Q = \frac{Q}{ND^3}$
	$x_2 + 1 = 0, \quad -3x_2 + z_2 - 1 = 0, \quad -y_2 - 1 = 0$
	$\therefore \quad x_2 = -1, \quad 3 + z_2 - 1 = 0, \quad y_2 = -1$
	$\therefore \quad z_2 = -2$
	$\therefore \quad \pi_2 = \rho^{-1} N^{-1} D^{-2} \mu = \frac{\mu}{\rho ND^2} = \frac{\gamma}{ND^2}$
	$\pi_3 = \rho^{x_3} N^{y_3} D^{x_3} P$
	$[M^0 L^0 T^0] = [M^{-3}]^{x_3}[T^{-1}]^{y_3}[L]^{z_3}[M^{-1}T^{-2}]$
	$\therefore f\left(\frac{Q}{ND^3}, \frac{gH}{N^2D^2}, \frac{v}{N^2}\right) = 0$ or constant
	$x_3 + 1 = 0, \quad -3x_3 + z_3 - 1 = 0, \quad -y_3 - 2 = 0$
	$\therefore \quad x_3 = -1, \quad 3 + z_3 - 1 = 0, \quad y_3 = -2$
	$\therefore \quad z_3 = -2$

	$\therefore \quad \pi_3 = \rho^{-1} N^{-2} D^{-2} p = \dfrac{P}{\rho N^2 D^2} = \dfrac{\rho \cdot g \cdot H}{\rho N^2 D^2} = \dfrac{gH}{N^2 D^2}$ $\therefore \quad \dfrac{Q}{ND^3} = f\left(\dfrac{gH}{N^2 D^2}, \dfrac{v}{N^2}\right)$ $\therefore \quad Q = N^3 \phi \left(\dfrac{gH}{N^2 D^2} \cdot \dfrac{v}{ND^2}\right)$

Ex. 6.10:	Drag on a body depends on its characteristic length l, speed u of the body Mass density and viscosity of the fluid and gravitational acceleration. Obtain an expression for drag in terms of dimensionless parameters. **Solution:** $f(F, l, u, \rho, \mu, g) = 0$ or constant Dimensions F l u ρ μ g $[MLT^{-2}]$ $[L]$ $[LT^{-1}]$ $[ML^{-3}]$ $[ML^{-1}T^{-1}]$ $[LT^{-2}]$ Taking ρ, l, u as repeating variables. $\pi_1 = \rho^{x_1} l_1^{y_1} u^2 F$ $[M^0 L^0 T^0] = [ML^{-3}]^{x_1} [LT^{y_1} [LT^{-1}]^{z_1} [MLT^{-2}]$ $x_1 + 1 = 0, \quad -3x_1 + y_1 + z_1 + 1 = 0, \quad -z_1 - 2 = 0$ $\therefore \quad x_1 = -1, \quad 3 + y_1 - 2 + 1 = 0, \quad z_1 = -2$ $\therefore \quad y_1 = -2$ $\therefore \quad \pi_1 = \rho^{-1} l^{-2} u^{-2} F = \dfrac{F}{\rho l^2 u^2}$ $\pi_2 = \rho^{x_2} l^{y_2} u^{z_2} \mu$

$$[M^0 L^0 T^0] = [ML^{-3}]^{x_2}[L]^{y_2}[LT^{-1}]^{z_2}[ML^{-1}T^{-1}]$$

$$-x_2 + 1 = 0, \quad -3x_2 + y_2 + z_2 - 1 = 0, \quad -z_2 - 1 = 0$$

$$\therefore \quad x_2 = -1, \quad 3 + y_2 - 1 - 1 = 0, \quad z_2 = -1$$

$$\therefore \quad y_2 = -1$$

$$\therefore \quad \pi_2 = \rho^{-1}\, l^{-1}\, u^{-1}\, \mu = \frac{\mu}{\rho u l} = \frac{1}{R_e};\ \frac{1}{\pi_2} = R_e \text{ (Reynold's number)}$$

$$\pi_3 = \rho^{x_3}\, l^{y_3} u^{z_1} g$$

$$[M^0 L^0 T^0] = [ML^{-3}]^{x_3}[L]^{y_3}[LT^{-1}]^2[LT^{-2}]$$

$$x_3 = 0, \quad -3x_3 + y_3 + z_3 + 1 = 0, \quad -z_3 - 2 = 0$$

$$\therefore \quad y_3 - 2 + 1 = 0, \quad z_3 = -2$$

$$\therefore \quad y_3 = 1$$

$$\therefore \quad \pi_3 = \rho^0\, l^1\, u^{-2}\, g = \frac{gl}{u^2}$$

$$\frac{1}{\pi_3} = \frac{u^2}{gl}, \quad \sqrt{\frac{1}{\pi_3}} = \frac{u}{\sqrt{gl}} = F_r \text{ (Froude number)}$$

$$\therefore\ f\left(\frac{F}{\rho R^2 u^2}, R_e, F_r\right) = 0 \text{ or constant}$$

$$\therefore \quad F = \rho R^2 u^2 \phi(R_e, F_r)$$

UNIT SUMMARY

Unit is summarized in the following points :

1. Dimensions of all quantities must be known correctly (Refer Table 6.3.1)
2. Rules for applying Buckingham π theorem.

3. Various non-dimensional parameters, like Froude number, Reynold's number, Mach number, Weber number, Euler number with their derivations.

6.7: Exercise

6.7.1: Objective Questions:

1. Which of the following is a primary quantity.

   a) Mass b) Density

   c) Speed d) Volume

   Ans: (d)

2. Which is the dimensionless quantity from the following?

   a) Mass b) Weight

   c) Specific weight d) Reynold's number

   Ans: (d)

3. Which of the following is the dimensions of force.

   a) $[M LT^{-1}]$ b) $[M LT^{-2}]$

   c) $[M L T^{-3}]$ d) $[M L^2 T^2]$

   Ans: (b)

4. The temperature is a primary quantity in dimensional analysis.

   a) True b) False

   Ans: (a)

5. The ratio of inertia force to the viscous force is called as.....

   a) Reynold's number b) Euler's number

   c) Weber's number d) None of the above

Ans: (a)

6. The dimensions of specific weight are -

a) $[M LT^{-1}]$ b) $[M LT^{-2}]$

c) $[M L T^{-3}]$ d) $[ML^{-2}T^{-2}]$

Ans: (d)

7. The reciprocal of Euler number is called as Newton number.

a) True b) False

Ans: (d)

8. The ratio of inertia force to the surface tension force is called as.....

a) Reynold's number b) Euler's number

c) Weber's number d) None of the above

Ans: (c)

9. Dimensional analysis converts large number of variables into

a) Larger number of dimensionless groups

b) Smaller number of dimensionless groups

c) Larger & smaller number of dimensionless groups

d) None

Ans: (b)

10. The size of a model in dimensional analysis is

a) > the prototype b) < the prototype

c) = the prototype d) None

Ans: (b)

11. First preferred repeating variable in Buckingham's Pi Theorem method is

a) Flow property b) Fluid property

c) Geometric property d) None

Ans: (c)

12. Dimensional homogeneity helps to determine the

a) Units of a physical quantity

b) Dimensions of a physical quantity

c) Units & dimensions of a physical quantity

d) None

Ans: (c)

13. Dimensional analysis develops which type of relations from the experimental data ?

a) Analytical b) Empirical

c) Analytical & empirical d) None

Ans: (b)

14. The dimensions of dynamic viscosity are

a) $[M LT^{-1}]$ b) $[M LT^{-2}]$

c) $[ML^{-1}T^{-1}]$ d) $[M L^2 T^2]$

Ans: (c)

15. Which is the dimensionless quantity?

a) Stress b) Strain

c) Modulus d) None

Ans: (b)

6.7.2: Theory Questions :

Q. 1 State Buckingham's π theorem.

Q. 2 Define and derive equations for following dimensionless numbers. Froude Number, Reynold's Number.

Q. 3 State dimensions of following variables.

(i) Angular velocity

(ii) Kinematic viscosity.

(iii) Moment of inertia of mass.

(iv) Momentum

Q. 4 Write short note on :

(i) Mach number

(ii) Euler number

(iii) Froude number

(iv) Reynolds number

Q.5. What do you understand by dimensional homogeneity ? Explain how dimensional analysis is used in analyzing fluid flow problems.

6.7. 3: Problems :

1. Using Buckingham's π-theorem, show that the velocity through a circular orifice is given by

$$V = \sqrt{2gH} f \left[\frac{D}{H}, \frac{\mu}{\rho \mathrm{VH}}\right]$$

H - Heading causing flow

2. The power required by an agitator in a tank is a function of the following variables: a. Diameter of the agitator b. Number of rotations of the impeller per unit time c. Viscosity of liquid d. Density of liquid

(i) From dimensional analysis using Buckingham's method, obtain a relation between power and the four variables.

(ii) The power consumption is found experimentally to be proportional to the square of the speed of rotation. By what factor would the power be expected to increase if the impeller diameter was doubled?

3. The efficiency of a fan η depends upon following factors :

   (i) Density ρ

   (ii) Dynamic viscosity of fluid μ

   (iii) Diameter of rotor D.

   (iv) Discharge ϕ

   (v) Angular velocity ω

   Show that

$$\eta = \phi\left(\frac{\mu}{D^2\omega\rho}, \frac{\phi}{d^3\omega}\right)$$

4. The discharge Q through an oil ring depends on the diameter d of oil ring, speed N rpm, mass density ρ of oil, absolute viscosity μ of oil, surface tension σ and specific weight γ of oil. Show that

$$Q = Nd^3\phi\left(\frac{\mu}{\rho N d^2}, \frac{\sigma}{\rho N^2 d^3}, \frac{\gamma}{\rho N^2 d}\right)$$

5. Torque T of a propeller depends upon density of liquid ρ, viscosity of liquid μ, theorem show that

$$T = \rho N^2 D^5 \phi\left[\frac{ND}{V_1}\frac{\rho N D^2}{\mu}\right]$$

QR CODES FOR SUPPORTING VIDEO

(1) (2) (3) (4) (5)

REFERENCES FOR FUTURE LEARNING

1. R.J. Garde and A.G. Mirajgaoker, Engineering Fluid Mechanics, SCITECH Publication, 2010.
2. Streeter & Wylie, Fluid Mechanics, Tata McGraw Hill, 9th edition.
3. K. Subramanya, Fluid Mechanics and Hydraulic Machines, Mc Graw hill, 2018.
4. Frank White, Fluid Mechanics, Mc Graw Hill, 7th edition.
5. J. Guthrie Brown, Hydro-Electric Engineering Practice, CBS Publishers & Distributors

CO AND PO ATTAINMENT TABLE

Following is the blank CO and PO attainment table, one can use this for individual practice for CO-PO mapping. Course outcomes (COs) for this course can be mapped with the programme outcomes (POs) after the completion of the course and a correlation can be made for the attainment of POs to analyze the gap. After proper analysis of the gap in the attainment of POs necessary measures can be taken to overcome the gaps.

Course Outcomes	**Expected Mapping with programme outcomes** *(1: Weak Correlation; 2: Medium Correlation; 3: Strong Correlation)*												
	PO-1	**PO-2**	**PO-3**	**PO-4**	**PO-5**	**PO-6**	**PO-6**	**PO-7**	**PO-8**	**PO-9**	**PO-10**	**PO-11**	**PO-12**
CO-1													
CO-2													
CO-3													
CO-4													
CO-5													

INDEX